Algebra 1

Larson
Boswell
Kanold
Stiff

Applications • Equations • Graphs

Chapter 12
Resource Book

The Resource Book contains the wide variety of blackline masters available for Chapter 12. The blacklines are organized by lesson. Included are support materials for the teacher as well as practice, activities, applications, and assessment resources.

McDougal Littell
A HOUGHTON MIFFLIN COMPANY
Evanston, Illinois • Boston • Dallas

Contributing Authors

The authors wish to thank the following individuals for their contributions to the Chapter 12 Resource Book.

Rita Browning
Linda E. Byrom
José Castro
Christine A. Hoover
Carolyn Huzinec
Karen Ostaffe
Jessica Pflueger
Barbara L. Power
Joanne Ricci
James G. Rutkowski
Michelle Strager

pages 33, 61, 76, 106: Excerpted and adapted from The World Book Encyclopedia. Copyright © 2000 World Book, Inc. By Permission of the publisher. www.worldbook.com

ISBN: 0-618-02050-0

15 14 13 12 11 10 -CKI- 06 05 04

Contents

12 Radicals and Connections to Geometry

Contents

Contents

Descriptions of Resources

This Chapter Resource Book is organized by lessons within the chapter in order to make your planning easier. The following materials are provided:

Tips for New Teachers These teaching notes provide both new and experienced teachers with useful teaching tips for each lesson, including tips about common errors and inclusion.

Parent Guide for Student Success This guide helps parents contribute to student success by providing an overview of the chapter along with questions and activities for parents and students to work on together.

Prerequisite Skills Review Worked-out examples are provided to review the prerequisite skills highlighted on the Study Guide page at the beginning of the chapter. Additional practice is included with each worked-out example.

Strategies for Reading Mathematics The first page teaches reading strategies to be applied to the current chapter and to later chapters. The second page is a visual glossary of key vocabulary.

Lesson Plans and Lesson Plans for Block Scheduling This planning template helps teachers select the materials they will use to teach each lesson from among the variety of materials available for the lesson. The block-scheduling version provides additional information about pacing.

Warm-Up Exercises and Daily Homework Quiz The warm-ups cover prerequisite skills that help prepare students for a given lesson. The quiz assesses students on the content of the previous lesson. (Transparencies also available)

Activity Support Masters These blackline masters make it easier for students to record their work on selected activities in the Student Edition.

Alternative Lesson Openers An engaging alternative for starting each lesson is provided from among these four types: *Application, Activity, Graphing Calculator,* or *Visual Approach.* (Color transparencies also available)

Graphing Calculator Activities with Keystrokes Keystrokes for four models of calculators are provided for each Technology Activity in the Student Edition, along with alternative Graphing Calculator Activities to begin selected lessons.

Practice A, B, and C These exercises offer additional practice for the material in each lesson, including application problems. There are three levels of practice for each lesson: A (basic), B (average), and C (advanced).

Contents

Reteaching with Practice These two pages provide additional instruction, worked-out examples, and practice exercises covering the key concepts and vocabulary in each lesson.

Quick Catch-Up for Absent Students This handy form makes it easy for teachers to let students who have been absent know what to do for homework and which activities or examples were covered in class.

Cooperative Learning Activities These enrichment activities apply the math taught in the lesson in an interesting way that lends itself to group work.

Interdisciplinary Applications/Real-Life Applications Students apply the mathematics covered in each lesson to solve an interesting interdisciplinary or real-life problem.

Math and History Applications This worksheet expands upon the Math and History feature in the Student Edition.

Challenge: Skills and Applications Teachers can use these exercises to enrich or extend each lesson.

Quizzes The quizzes can be used to assess student progress on two or three lessons.

Chapter Review Games and Activities This worksheet offers fun practice at the end of the chapter and provides an alternative way to review the chapter content in preparation for the Chapter Test.

Chapter Tests A, B, and C These are tests that cover the most important skills taught in the chapter. There are three levels of test: A (basic), B (average), and C (advanced).

SAT/ACT Chapter Test This test also covers the most important skills taught in the chapter, but questions are in multiple-choice and quantitative-comparison format. (See *Alternative Assessment* for multi-step problems.)

Alternative Assessment with Rubrics and Math Journal A journal exercise has students write about the mathematics in the chapter. A multi-step problem has students apply a variety of skills from the chapter and explain their reasoning. Solutions and a 4-point rubric are included.

Project with Rubric The project allows students to delve more deeply into a problem that applies the mathematics of the chapter. Teacher's notes and a 4-point rubric are included.

Cumulative Review These practice pages help students maintain skills from the current chapter and preceding chapters.

Algebra 1
Chapter 12 Resource Book

Tips for New Teachers

For use with Chapter 12

LESSON 12.1

TEACHING TIP Although at this level, students do not know that a root is a power with a rational exponent, they need to know how roots fit into the order of operations. Review *order of operations* and ask students where they would place square roots in it. You can write a new order of operations where the step after taking care of grouping symbols is to evaluate powers and/or square roots.

COMMON ERROR Show students a graph of the form $y = \sqrt{x} + k$ to illustrate that these graphs do not always start on the *x*-axis. Completing the activity on page 715 can help students become more familiar with graphs of square-root functions.

TEACHING TIP After students practice graphing square-root functions, you might want to introduce the idea of a *family of functions*. Once students understand that all functions in a family have graphs of the same shape, ask them what families of functions they know. Make a list of different families of functions and review how to tell whether a function belongs to one family or another. Students should also be able to quickly sketch the graphs of these functions based on what they have already learned.

LESSON 12.2

COMMON ERROR Some students combine radicals by adding or subtracting both the coefficients and the radicands. For example, they might evaluate $3\sqrt{5} + 6\sqrt{2}$ as $9\sqrt{7}$. Ask students to evaluate both expressions with their calculator to show that they are not equal. Then, ask your students whether they remember how to combine like terms and have one of them explain how to do so.

COMMON ERROR Students might multiply or simplify radical expressions by incorrectly mixing the coefficients with the radicands. For instance, they might think

$$\frac{\sqrt{6}}{3} = \frac{\sqrt{3 \cdot 2}}{3} = \sqrt{2} \text{ or } 2 \cdot \sqrt{8} = \sqrt{16} = 4.$$

Again, ask your students to evaluate these expressions with their calculators to check whether they are correct. If you feel that the words

radicand and *coefficient* are confusing students, tell them that they can multiply or simplify things *inside the radical* and things *outside the radical*, but they cannot mix them.

INCLUSION Even after reviewing the patterns for a *square of a binomial* and *sum and difference* some students might make mistakes using them. If that is the case, review FOIL with them and give them the option to use either method to find the answer for that type of exercise.

LESSON 12.3

COMMON ERROR When solving radical equations, some students always solve for $\sqrt{x}$, regardless of what the radicand is. To solve an equation such as $\sqrt{x - 3} = 1$, they might add 3 to each side of the equation, resulting in $\sqrt{x} = 4$. Remind your students that they need to isolate the radical with everything inside it. To help them with these problems, ask them to circle the radical in the equation and then to isolate that circled expression.

COMMON ERROR Students who did not understand the difference between $\sqrt{}$, $-\sqrt{}$, and $\pm\sqrt{}$ will have a hard time identifying extraneous solutions. They might argue that 169 is a valid solution for $\sqrt{x} + 13 = 0$ because "the square root of 169 can be either positive 13 or negative 13." Ask your students to identify what type of root the problem is asking for, positive or negative, before they check the validity of the solutions.

LESSON 12.4

INCLUSION Completing the square can be an abstract procedure where each step is totally meaningless. The activity on page 729 will provide your students with a concrete model that shows the reason behind each step taken as well as the name for the procedure.

TEACHING TIP This is a great moment to review and summarize the five methods that students have learned to solve quadratic equations. For each method, ask your students to write pros and cons as well as an example of an equation that could easily be solved using that specific method. You can have students solve some of their own equations to review each method.

LESSON 12.5

TEACHING TIP The Pythagorean theorem can lead some students to believe that if a theorem is true its converse must also be true. Provide students with an example of a conditional statement whose converse is not true. As an enrichment activity, have students work with real-life conditional statements and their converses. Write them in "if-then" form and analyze whether the statement or its converse is true.

LESSON 12.6

COMMON ERROR If your students subtract coordinates instead of adding them, ask students to plot the endpoints of the segment and the midpoint they found. Their drawing will show them that they did not actually find the midpoint.

LESSON 12.7

TEACHING TIP Use the mnemonic trick known as "SOHCAHTOA" to help your students to remember the trigonometric ratios. "SOHCAHTOA" is just the initial letter of each word in the ratios, as shown below:

$$\underline{S}ine = \frac{\underline{O}pposite}{\underline{H}ypotenuse} \text{ (SOH)}$$

$$\underline{C}osine = \frac{\underline{A}djacent}{\underline{H}ypotenuse} \text{ (CAH)}$$

$$\underline{T}angent = \frac{\underline{O}pposite}{\underline{A}djacent} \text{ (TOA)}$$

COMMON ERROR Since the hypotenuse of a right triangle is always the same side regardless of the angle being considered, some students think that the same must be true for the opposite and adjacent sides. Make sure to complete a problem such as Example 1 on page 752 where students have to find the trigonometric ratios for the two non-right angles of a right triangle. This problem requires students to correctly define the opposite and adjacent side for each of these two angles.

TEACHING TIP You can have your students investigate the relationships between the trigonometric ratios of the two non-right angles of a right triangle. If the angles are labeled as $\angle A$ and $\angle B$, students will discover the following:

$$\sin A = \cos B, \cos A = \sin B, \tan A = \frac{1}{\tan B}.$$

LESSON 12.8

TEACHING TIP Students do not always see a need to prove a theorem. Students at this level are usually satisfied by testing a few concrete examples to check the validity of a theorem. To show students that this is not enough, give them a statement such as "all numbers squared yield a larger or equal number." Ask your students whether this is a true statement and challenge them to test it with all kinds of numbers—positive and negative numbers, even zero. Most students will say that the statement is true, because it always works with integers. However, the statement is false as students should quickly discover if you challenge them to test it with rational numbers.

Outside Resources

BOOKS/PERIODICALS

Crites, Terry W. "Connecting Geometry and Algebra: Geometric Interpretations of Distance." *Mathematics Teacher* (April 1995); pp. 292–297.

Miller, William and Linda Wagner. "Pythagorean Dissection Puzzles." *Mathematics Teacher* (April 1993); pp. 302–314.

ACTIVITIES/MANIPULATIVES

Peterson, Blake E., Patrick Averbeck, and Lynanna Baker. "Sine Curves and Spaghetti." *Activities: Mathematics Teacher* (October 1998); pp. 564–567.

SOFTWARE

Harvey, Wayne, Judah Schwartz, and Michael Yerushalmy. *Geometric superSupposer.* Scotts Valley, CA; Sunburst.

NAME _____ DATE _____

Parent Guide for Student Success

For use with Chapter 12

Chapter Overview One way that you can help your student succeed in Chapter 12 is by discussing the lesson goals in the chart below. When a lesson is completed, ask your student to interpret the lesson goals for you and to explain how the mathematics of the lesson relates to one of the key applications listed in the chart.

Lesson Title	Lesson Goals	Key Applications
12.1: Square-Root Functions	Evaluate and graph a square-root function. Use square-root functions to model real-life problems.	• Dinosaurs • Medicine • Investigating Accidents
12.2: Operations with Radical Expressions	Add, subtract, multiply, and divide radical expressions. Use radical expressions in real-life situations.	• Sailing • Pole-Vaulting • Falling Objects
12.3: Solving Radical Equations	Solve a radical equation. Use radical equations to solve real-life problems.	• Plane De-icing • Amusement Park Ride • Blotting Paper
12.4: Completing the Square	Solve a quadratic equation by completing the square. Choose a method for solving a quadratic equation.	• Vettisfoss Falls • Diving • Penguins
12.5: The Pythagorean Theorem and Its Converse	Use the Pythagorean Theorem and its converse in real-life problems.	• Baseball Diamond • Designing a Staircase • Planting a New Tree
12.6: The Distance and Midpoint Formulas	Find the distance between two points and the midpoint between two points in a coordinate plane.	• Soccer • Computer Games • Distances on Maps
12.7: Trigonometric Ratios	Use the sine, cosine, and tangent of an angle. Use trigonometric ratios in real-life problems.	• Parasailing • Operating a Crane • Cloud Height
12.8: Logical Reasoning: Proof	Use logical reasoning and proof to prove a statement is true. Prove that a statement is false.	• Goldbach's Conjecture • Lawyers • Four-Color Problem

Test-Taking Strategy

*On multiple choice tests, try to **make an educated guess** when you are unsure of the answer. First **eliminate answers** that you know cannot be correct.* Have your student look at the multiple choice test on pages 770-771. Ask your student to explain how to eliminate some answers to one of the questions without actually finding the correct answer.

NAME _____ DATE _____

Parent Guide for Student Success

For use with Chapter 12

Key Ideas Your student can demonstrate understanding of key concepts by working through the following exercises with you.

Lesson	Exercise
12.1	The radius of a cylinder is given by $r = \sqrt{\dfrac{V}{\pi h}}$, where V is the volume of the cylinder and h is its height. Find the radius of a can that holds 330 cubic centimeters of a soft drink and has a height of 11.5 centimeters. Use 3.14 as an approximation for π.
12.2	Simplify the expression. $\dfrac{\sqrt{6}}{\sqrt{2-1}}$
12.3	The speed s (in meters per second) of a fast-moving ocean wave called a tsunami is given by $s = \sqrt{gd}$, where g is 9.8 meters per second squared and d is the depth of the ocean. If a tsunami is moving at a speed of 140 meters per second, what is the depth of the ocean?
12.4	Solve the equation by completing the square: $3x^2 - 6x - 10 = 2$.
12.5	In order to get a roof line 15 feet long on each side, a carpenter makes the gable 9 feet high and 24 feet long. Why does this work?
12.6	Find the midpoint of the segment between $(3, -4)$ and $(-5, 2)$. Find the length of the segment.
12.7	Standing 60 feet from a tree, you measure the angle between the ground and your line of sight to the top of the tree as 30°. What is the height of the tree?
12.8	Find a counterexample to show the statement is *not true*: $ab > b$.

Home Involvement Activity

You will need: A protractor, string, a small heavy object

Directions: Tie the heavy object to one end of the string. Tie the other end of the string to the hole near 0 on the protractor. Hold the protractor upside down with the straight edge on top. Stand back and sight the top of your home along the straight edge of the protractor. Have your student read the angle (less than 90°) where the string falls on the protractor. Subtract this angle from 90°. The result is the angle between your line of sight and the ground. Measure the distance from your position to the base of the building. Use trigonometric ratios to find the height of your home.

Answers

12.1: about 3 cm **12.2:** $2\sqrt{3} + \sqrt{6}$ **12.3:** 2000 m **12.4:** $1 \pm \sqrt{5}$ **12.5:** The gable has 2 right triangles, each 9 ft by 12 ft by 15 ft and $9^2 + 12^2 = 15^2$ **12.6:** $(-1, -1)$; 10 **12.7:** about 34.6 ft **12.8** *Sample answer:* $\frac{1}{2}(6) < 6$

NAME _____ DATE _____

Prerequisite Skills Review

For use before Chapter 12

EXAMPLE 1 *Solving Problems with Similar Triangles*

The following two triangles are similar. Find the length of the side marked *x*.

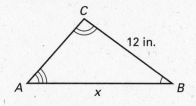

SOLUTION

Set up equal ratios to find the unknown side length.

$$\frac{\text{Length of } AB}{\text{Length of } DE} = \frac{\text{Length of } BC}{\text{Length of } EF}$$

Length of $AB = x$

Length of $DE = 8$

Length of $BC = 12$

Length of $EF = 6$

$\dfrac{x}{8} = \dfrac{12}{6}$ Write algebraic model.

$24\left(\dfrac{x}{8}\right) = 24\left(\dfrac{12}{6}\right)$ Multiply each side by 24.

$3x = 48$ Simplify.

$x = 16$ Divide both sides by 3.

The length of *AB* is 16 inches.

Exercises for Example 1

In Exercises 1–4, the two triangles are similar. Find the length of the side marked *x*.

1.

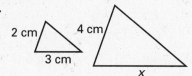

2.

3.

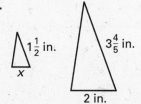

4.

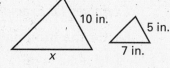

Prerequisite Skills Review

For use before Chapter 12

EXAMPLE 2 *Graphing an Equation*

Simplify the expression.

$$\sqrt{\frac{12}{243}}$$

SOLUTION

$$\sqrt{\frac{12}{243}} = \sqrt{\frac{3 \cdot 4}{3 \cdot 81}} \quad \text{Divide out common factors.}$$

$$= \sqrt{\frac{4}{81}} \quad \text{Simplify.}$$

$$= \frac{\sqrt{4}}{\sqrt{81}} \quad \text{Use quotient property.}$$

$$= \frac{2}{9}$$

Exercises for Example 2

Simplify the expression.

5. $\sqrt{\dfrac{5}{144}}$ **6.** $\dfrac{\sqrt{96}}{8}$ **7.** $\dfrac{\sqrt{48}}{\sqrt{100}}$ **8.** $\sqrt{\dfrac{8}{40}}$

EXAMPLE 3 *Checking solutions of a Linear Inequality*

Factor the trinomial.

$x^2 + 7x - 120$

SOLUTION

For this trinomial, $b = 7$ and $c = -120$. Because c is negative, you
know that p and q cannot both have negative values.

$$x^2 + 7x - 120 = (x + p)(x + q) \qquad \text{Find } p \text{ and } q \text{ values when } p + q = 7 \text{ and}$$
$$pq = -120.$$

$$= (x + 15)(x - 8) \qquad p = 15 \text{ and } q = -8$$

CHECK: Use a graphing calculator. Graph $x^2 + 7x - 120$ and $(x + 15)(x - 8)$ on the
same screen.

The graphs coincide, so your answer is correct.

Exercises for Example 3

Factor the trinomial.

9. $x^2 + 8x - 20$ **10.** $x^2 + 20x + 64$ **11.** $x^2 - 17x + 72$

NAME _____ DATE _____

Strategies for Reading Mathematics

For use with Chapter 12

Strategy: Reading Diagrams

Diagrams are an effective way to communicate information because they show us, rather than tell us, what we need to know. Diagrams convey a lot of information in a small space. Pay attention to the details and don't assume anything from a diagram that is not explicitly indicated.

You can tell a lot about the triangles below just by looking at the diagrams. For example, you can tell that the triangle at the left is a right triangle and that the legs of the triangle are the same length. You can also tell that P is the midpoint of the hypotenuse, because the segments on either side of P are the same length. Since no actual lengths are indicated, you can't assume anything about the lengths of the sides except that the hypotenuse is longer than each leg.

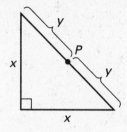   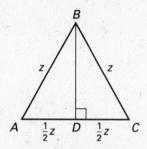

STUDY TIP
Reading Diagrams

Sometimes diagrams illustrate explanations. Try to match the parts of the explanation to the diagram. As you study a diagram, ask yourself questions, such as *How is this diagram the same as others I have seen? How is it different?*

STUDY TIP
Draw Diagrams

One way to make yourself a better reader of diagrams is to draw diagrams yourself. As you read about new concepts or solve problems, make sketches to help you see what you are reading about. Show all the details and information that you can.

Questions

Use the triangles above for Questions 1–4.

1. Are there any right triangles in the figure at the right? If so, name them using the lettered vertices. (For example, the large triangle can be named triangle *ABC*.)

2. What additional information can you get from the figure at the right?

3. The Pythagorean theorem applies only to right triangles. How can you tell from a sketch of a triangle if the Pythagorean theorem applies to the triangle?

4. A square has four right angles and four sides that are equal in length. Draw a diagram of a square and label it so that anyone looking at your sketch will know that the figure is a square.

NAME _____ DATE _____

Strategies for Reading Mathematics

For use with Chapter 12

Visual Glossary

The Study Guide on page 708 lists the key vocabulary for Chapter 12 as well as review vocabulary from previous chapters. Use the page references on page 708 or the Glossary in the textbook to review key terms from prior chapters. Use the visual glossary below to help you understand some of the key vocabulary in Chapter 12. You may want to copy these diagrams into your notebook and refer to them as you complete the chapter.

GLOSSARY

square-root function (p. 709) A function defined by the equation $y = \sqrt{x}$.

hypotenuse (p. 738) The side opposite the right angle in a right triangle.

legs of a right triangle (p. 738) The two sides of a right triangle that are not opposite the right angle.

Pythagorean Theorem (p. 738) If a triangle is a right triangle, then the sum of the squares of the lengths of the legs a and b equals the square of the length of the hypotenuse c.

trigonometric ratio (p. 752) The ratio of the lengths of two sides of a right triangle. The cosine, sine, and tangent ratios.

Square-Root Functions

The radicand in a square-root function must always be a number greater than or equal to zero.

square-root functions

$$y = \sqrt{x}$$
$$x \geq 0$$

$$y = \sqrt{x+2}$$
$$x+2 \geq 0$$
$$x \geq -2$$

$$y = \sqrt{x-1}$$
$$x-1 \geq 0$$
$$x \geq 1$$

Right Triangles

The triangle is a right triangle.

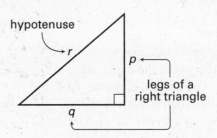

$$p^2 + q^2 = r^2 \longleftarrow \text{Pythagorean Theorem}$$

If a triangle is a right triangle, then the sum of the squares of the lengths of the legs (p and q) equals the square of the length of the hypotenuse (r).

Trigonometric Ratios

The trigonometric ratios for $\angle D$ in the triangle are determined by the lengths of the sides.

$$\sin D = \frac{\text{opposite}}{\text{hypotenuse}} = \frac{d}{e}$$

$$\cos D = \frac{\text{adjacent}}{\text{hypotenuse}} = \frac{f}{e}$$

$$\tan D = \frac{\text{opposite}}{\text{adjacent}} = \frac{d}{f}$$

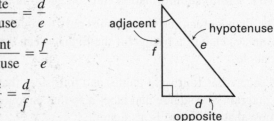

TEACHER'S NAME _____ CLASS _____ ROOM _____ DATE _____

Lesson Plan

1-day lesson (See *Pacing the Chapter,* TE pages 706C–706D)   **For use with pages 709–715**

GOALS
1. **Evaluate and graph a square-root function.**
2. **Use square-root functions to model real-life problems.**

State/Local Objectives _____

_____

✓ **Check the items you wish to use for this lesson.**

STARTING OPTIONS
_____ Prerequisite Skills Review: CRB pages 5–6
_____ Strategies for Reading Mathematics: CRB pages 7–8
_____ Warm-Up or Daily Homework Quiz: TE pages 709 and 697, CRB page 11, or Transparencies

TEACHING OPTIONS
_____ Motivating the Lesson: TE page 710
_____ Lesson Opener (Application): CRB page 12 or Transparencies
_____ Graphing Calculator Activity with Keystrokes: CRB page 13
_____ Examples: 1–4: SE pages 709–711
_____ Extra Examples: TE pages 710–711 or Transparencies; Internet
_____ Technology Activity: SE page 715
_____ Closure Question: TE page 711
_____ Guided Practice Exercises: SE page 712

APPLY/HOMEWORK
Homework Assignment
_____ Basic   20–56 even, 57, 60, 63, 65, 70, 75, 80, 86, 87
_____ Average   20–56 even, 57, 60, 63, 65, 70, 75, 80, 86, 87
_____ Advanced   20–56 even, 57, 60, 63–65, 70, 75, 80, 86, 87

Reteaching the Lesson
_____ Practice Masters: CRB pages 14–16 (Level A, Level B, Level C)
_____ Reteaching with Practice: CRB pages 17–18 or Practice Workbook with Examples
_____ Personal Student Tutor

Extending the Lesson
_____ Applications (Interdisciplinary): CRB page 20
_____ Challenge: SE page 714; CRB page 21 or Internet

ASSESSMENT OPTIONS
_____ Checkpoint Exercises: TE pages 710–711 or Transparencies
_____ Daily Homework Quiz (12.1): TE page 714, CRB page 24, or Transparencies
_____ Standardized Test Practice: SE page 714; TE page 714; STP Workbook; Transparencies

Notes _____

_____

_____

Lesson Plan for Block Scheduling

Half-day lesson (See *Pacing the Chapter,* TE pages 706C–706D)   For use with pages 709–715

GOALS 1. **Evaluate and graph a square-root function.**
2. **Use square-root functions to model real-life problems.**

State/Local Objectives _____

_____

_____

✓ **Check the items you wish to use for this lesson.**

STARTING OPTIONS
____ Prerequisite Skills Review: CRB pages 5–6
____ Strategies for Reading Mathematics: CRB pages 7–8
____ Warm-Up or Daily Homework Quiz: TE pages 709 and
        697, CRB page 11, or Transparencies

TEACHING OPTIONS
____ Motivating the Lesson: TE page 710
____ Lesson Opener (Application): CRB page 12 or Transparencies
____ Graphing Calculator Activity with Keystrokes: CRB page 13
____ Examples 1–4: SE pages 709–711
____ Extra Examples: TE pages 710–711 or Transparencies; Internet
____ Technology Activity: SE page 715
____ Closure Question: TE page 711
____ Guided Practice Exercises: SE page 712

APPLY/HOMEWORK
Homework Assignment (See also the assignment for Lesson 12.2.)
____ Block Schedule: 20–56 even, 57, 60, 63, 65, 70, 75, 80, 86, 87

Reteaching the Lesson
____ Practice Masters: CRB pages 14–16 (Level A, Level B, Level C)
____ Reteaching with Practice: CRB pages 17–18 or Practice Workbook with Examples
____ Personal Student Tutor

Extending the Lesson
____ Applications (Interdisciplinary): CRB page 20
____ Challenge: SE page 714; CRB page 21 or Internet

ASSESSMENT OPTIONS
____ Checkpoint Exercises: TE pages 710–711 or Transparencies
____ Daily Homework Quiz (12.1): TE page 714, CRB page 24, or Transparencies
____ Standardized Test Practice: SE page 714; TE page 714; STP Workbook; Transparencies

Notes _____

_____

_____

CHAPTER PACING GUIDE	
Day	**Lesson**
1	**12.1 (all)**; 12.2 (begin)
2	12.2 (end); 12.3 (begin)
3	12.3 (end); 12.4 (begin)
4	12.4 (end); 12.5 (begin)
5	12.5 (end); 12.6 (all)
6	12.7 (all); 12.8 (all)
7	Review/Assess Ch. 12

NAME _____ DATE _____

WARM-UP EXERCISES

For use before Lesson 12.1, pages 709–715

Find the value of each expression to the nearest tenth when $x = 3$.

1. $\sqrt{x}$

2. $\sqrt{x + 1}$

3. $\sqrt{x^3 - 2}$

4. Which of the following is not a real number?

    a. $\sqrt{-9}$ **b.** $\sqrt{144}$ **c.** $\sqrt{18.5}$

DAILY HOMEWORK QUIZ

For use after Lesson 11.8 pages 690–698

1. Solve $\dfrac{4}{2x - 1} = \dfrac{3}{x - 6}$ by cross multiplying.

Solve the equation.

2. $\dfrac{3}{x} + \dfrac{2}{3} = \dfrac{1}{3x}$ **3.** $1 + \dfrac{x - 1}{x + 8} = -\dfrac{15}{x^2 + 3x - 40}$

4. $\dfrac{x}{x - 4} + \dfrac{3x}{x - 6} = \dfrac{x^2 - 5x - 4}{x^2 - 10x + 24}$

5. Sketch a graph of $y = \dfrac{3}{x - 2} + 1$. Describe the domain.

6. After 8 games of bowling in your league, your average score is 145. If you could score 160 on your next few games, how many games would it take to raise your average to 150?

Application Lesson Opener

For use with pages 709–714

1. Each side of a square playground is 15 feet long. If x represents the area of the playground, which equation can be used to model this situation? Explain.

A. $15 = 4x$ **B.** $\dfrac{x}{4} = 15$

C. $15 = \sqrt{x}$ **D.** $x = \sqrt{15}$

2. Let y represent the length of each side of a square playground and x represent the area. Which equation can be used to model this situation? Explain.

A. $y = 4x$ **B.** $\dfrac{x}{4} = y$

C. $y = \sqrt{x}$ **D.** $x = \sqrt{y}$

3. You are given the area, x, of a square. Use the equation you found in Question 2 to help you find the length of a side, y, of the square.

  a. area = 9   length = _____

  b. area = 16   length = _____

  c. area = 25   length = _____

  d. area = 36   length = _____

  e. area = 49   length = _____

  f. area = 100   length = _____

4. If you know the area of a square, how can you find the length of a side?

Graphing Calculator Activity Keystrokes

For use with Technology Activity 12.1 on page 715

TI-82

Y= | 2nd | [√] | X,T,θ | ENTER

2 | 2nd | [√] | X,T,θ | ENTER

3 | 2nd | [√] | X,T,θ | ENTER

4 | 2nd | [√] | X,T,θ | ENTER

ZOOM | 6

Y= | CLEAR | (-) | 2nd | [√] | X,T,θ | ENTER

CLEAR | 2nd | [√] | X,T,θ | ENTER | CLEAR

ENTER | CLEAR

GRAPH

Y= | CLEAR | (-) | 4 | 2nd | [√] | X,T,θ | ENTER

CLEAR | 4 | 2nd | [√] | X,T,θ | ENTER

GRAPH

Y= | CLEAR | (-) | 9 | 2nd | [√] | X,T,θ | ENTER

CLEAR | 9 | 2nd | [√] | X,T,θ | ENTER

GRAPH

TI-83

Y= | 2nd | [√] | X,T,θ,n |) | ENTER

2 | 2nd | [√] | X,T,θ,n |) | ENTER

3 | 2nd | [√] | X,T,θ,n |) | ENTER

4 | 2nd | [√] | X,T,θ,n |) | ENTER

ZOOM | 6

Y= | CLEAR | (-) | 2nd | [√] | X,T,θ,n | ENTER

CLEAR

2nd | [√] | X,T,θ,n | ENTER | CLEAR | ENTER

CLEAR

GRAPH

Y= | CLEAR | (-) | 4 | 2nd | [√] | X,T,θ,n | ENTER

CLEAR | 4 | 2nd | [√] | X,T,θ,n | ENTER

GRAPH

Y= | CLEAR | (-) | 9 | 2nd | [√] | X,T,θ,n | ENTER

CLEAR | 9 | 2nd | [√] | X,T,θ,n | ENTER

GRAPH

SHARP EL-9600c

Y= | 2ndF | [√] | X/θ/T/n | ENTER

2 | 2ndF | [√] | X/θ/T/n | ENTER

3 | 2ndF | [√] | X/θ/T/n | ENTER

4 | 2ndF | [√] | X/θ/T/n | ENTER

ZOOM | [A] 5

Y= | CL | (-) | 2ndF | [√] | X/θ/T/n | ENTER

CL | 2ndF | [√] | X/θ/T/n | ENTER | CL

ENTER | CL | GRAPH

Y= | CL | (-) | 4 | 2ndF | [√] | X/θ/T/n | ENTER

CL | 4

2ndF | [√] | X/θ/T/n | ENTER

GRAPH

Y= | CL | (-) | 9 | 2ndF | [√] | X/θ/T/n | ENTER

CL | 9 | 2ndF | [√] | X/θ/T/n | ENTER

GRAPH

CASIO CFX-9850Ga PLUS

From the main menu, choose GRAPH.

SHIFT | [√] | X,θ,T | EXE

2 | SHIFT | [√] | X,θ,T | EXE

3 | SHIFT | [√] | X,θ,T | EXE

4 | SHIFT | [√] | X,θ,T | EXE

SHIFT | F3 | F3 | EXIT | F6

EXIT | ▲ | ▲ | ▲ | (-) | SHIFT | [√] | X,θ,T | EXE

SHIFT | [√] | X,θ,T | EXE | F2 | F1 | ▼ | F2 | F1

F6 | (-) | 4 | SHIFT | [√] | X,θ,T | EXE | 4 | SHIFT | [√]

X,θ,T | EXE | F6

EXIT | ▲ | ▲ | (-) | 9 | SHIFT | [√] | X,θ,T | EXE

9 | SHIFT | [√] | X,θ,T | EXE | F6

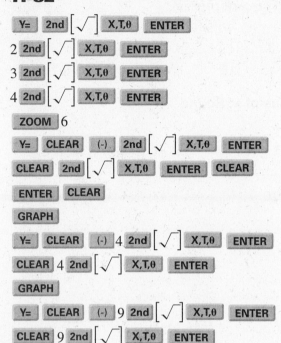

Practice A

For use with pages 709–714

State the two whole numbers between which each square root is located.

1. $\sqrt{12}$ 2. $\sqrt{48}$ 3. $\sqrt{92}$ 4. $\sqrt{26}$

5. $\sqrt{75}$ 6. $\sqrt{54}$ 7. $\sqrt{135}$ 8. $\sqrt{120}$

Evaluate the square-root function for the given value of *x*. Round your answer to the nearest tenth.

9. $y = 4\sqrt{x}$; 25

10. $y = \frac{1}{2}\sqrt{x + 7}$; 2

11. $y = \sqrt{3x + 2}$; 7

12. $y = \sqrt{25 - 2x}$; 3

13. $y = \sqrt{\frac{x}{2} + 20}$; 9

14. $y = 3\sqrt{\frac{2x}{3} - 7}$; 21

Find the domain of the function.

15. $y = 7\sqrt{x}$

16. $y = \sqrt{x - 4}$

17. $y = \sqrt{x + 8}$

18. $y = 3 - \sqrt{x + 2}$

19. $y = \frac{\sqrt{x}}{4}$

20. $y = \frac{3}{4}\sqrt{2x + 1}$

21. $y = \sqrt{4x + 4}$

22. $y = \sqrt{3x + 9}$

23. $y = \sqrt{5x + 3}$

Match the function with its graph.

A.
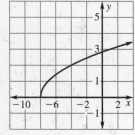

B.
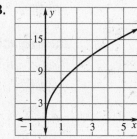

C.
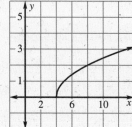

24. $y = 7\sqrt{x}$

25. $y = \sqrt{x - 4}$

26. $y = \sqrt{x + 8}$

27. *Medicine*   A doctor may need to know a person's body surface area to prescribe the correct amount of medicine. You can use the model below to find a person's body surface area *A* (in square meters) where *h* represents height (in inches) and *w* represents weight (in pounds).

Body surface area model: $A = \sqrt{\dfrac{h \cdot w}{3131}}$

Find the body surface area of a person who is 5 feet 6 inches tall and weighs 120 pounds. Round your answer to the nearest hundredth.

NAME _____ DATE _____

Practice B

For use with pages 709–714

Evaluate the square-root function for the given value of x. Round your answer to the nearest tenth.

1. $y = 3\sqrt{x};\ 36$

2. $y = \frac{1}{3}\sqrt{x + 6};\ 3$

3. $y = \sqrt{2x + 5};\ 12$

4. $y = \sqrt{36 - 3x};\ 7$

5. $y = \sqrt{\frac{x}{2} + 42};\ 16$

6. $y = 4\sqrt{\frac{3x}{4} - 3};\ 28$

7. $y = \sqrt{70 - 3x};\ 15$

8. $y = \sqrt{\frac{x}{2} + 48};\ 16$

9. $y = 10\sqrt{\frac{x}{5} - 3};\ 65$

Find the domain of the function.

10. $y = -5\sqrt{x}$

11. $y = \sqrt{x - 9}$

12. $y = \sqrt{x + 4.5}$

13. $y = 4x\sqrt{2x}$

14. $y = \sqrt{x - 7}$

15. $y = 0.1\sqrt{x + 6}$

16. $y = 6 - \sqrt{6x + 3}$

17. $y = \frac{\sqrt{2x}}{8}$

18. $y = -\frac{3}{4}\sqrt{3x - 4}$

19. $y = \sqrt{\frac{4}{5}x}$

20. $y = \sqrt{7x - 2}$

21. $y = \sqrt{15x - 5}$

Match the function with its graph.

A.
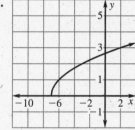

B.
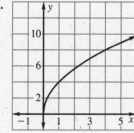

C.
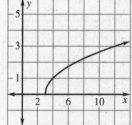

22. $y = 4\sqrt{x}$

23. $y = \sqrt{x - 3}$

24. $y = \sqrt{x + 7}$

Find the domain and the range of the function. Then sketch the graph of the function.

25. $y = 2\sqrt{x}$

26. $y = \sqrt{x} + 6$

27. $y = \sqrt{x} - 5$

28. $y = \sqrt{2x + 1}$

29. $y = \sqrt{x + 6}$

30. $y = \sqrt{3x - 12}$

31. *Geometry* Find the radius of a circle if the area is 54 square inches. Use the model below where *A* represents the area (in square inches) and *r* represents the radius (in inches).

$$\text{Radius of a circle model: } r = \sqrt{\frac{A}{\pi}}$$

NAME _____ DATE _____

Practice C

For use with pages 709–714

Evaluate the square-root function for the given value of x. Round your answer to the nearest tenth.

1. $y = 8\sqrt{x}$; 49

2. $y = \frac{1}{3}\sqrt{x + 10}$; 15

3. $y = \sqrt{7x + 15}$; 8

4. $y = \sqrt{56 - 4x}$; 8

5. $y = \sqrt{\frac{x}{5} + 72}$; 35

6. $y = 9\sqrt{\frac{4x}{5} - 12}$; 55

7. $y = \sqrt{62 - 4x}$; 12

8. $y = \sqrt{\frac{x}{3} + 36}$; 36

9. $y = 4\sqrt{x^2 - 7}$; 8

10. $y = 3\sqrt{x^2 + 4}$; 8

11. $y = \sqrt{9x - 9}$; 6

12. $y = \sqrt{3x + \frac{x}{6}}$; 12

Find the domain of the function.

13. $y = 9\sqrt{7x}$

14. $y = \sqrt{x - 3}$

15. $y = 0.2\sqrt{x + 6.5}$

16. $y = 6 - \sqrt{8x + 6}$

17. $y = -\frac{\sqrt{4x}}{5}$

18. $y = -\frac{2}{5}\sqrt{5x - 2}$

19. $y = \sqrt{\frac{3}{10}x}$

20. $y = \sqrt{9x - 4}$

21. $y = 0.1\sqrt{18x - 3}$

22. $y = \sqrt{\frac{4}{5}x + 8}$

23. $y = \sqrt{4(x - 2)}$

24. $y = \sqrt{x^2}$

Find the domain and the range of the function. Then sketch the graph of the function.

25. $y = 4\sqrt{x}$

26. $y = \sqrt{x + 9}$

27. $y = \sqrt{x} - 2.5$

28. $y = \sqrt{5x + 2}$

29. $y = \sqrt{3x - 1}$

30. $y = \sqrt{6x - 2}$

31. $y = \sqrt{2x + 3}$

32. $y = \sqrt{x + 2} - 3$

33. $y = \sqrt{x - 5} + 1$

Fire Hoses   **In Exercises 34–36, use the following information.**

For a fire hose with a nozzle that has a diameter of 2 inches, the flow rate f (in gallons per minute) can be modeled by $f = 120\sqrt{p}$ where p is the nozzle pressure in pounds per square inch.

34. Sketch a graph of the model. Label the units on the axes.

35. If the flow rate is 800 gallons per minute, what is the nozzle pressure?

36. If the flow rate is doubled to 1600 gallons per minute, does the nozzle pressure double? Explain.

Reteaching with Practice

For use with pages 709–714

GOAL Evaluate and graph a square-root function and use square-root functions to model real-life problems

---

> **VOCABULARY**
>
> A **square-root function** is defined by the equation $y = \sqrt{x}$.

EXAMPLE 1 *Graphing $y = a\sqrt{x} + k$*

Find the domain and the range of $y = 3\sqrt{x} + 2$. Then sketch its graph.

SOLUTION

The domain is the set of all nonnegative numbers. The range is the set of all numbers that are greater than or equal to 2. Make a table of values, plot the points, and connect them with a smooth curve.

x	y
0	$y = 3\sqrt{0} + 2 = 2$
1	$y = 3\sqrt{1} + 2 = 5$
2	$y = 3\sqrt{2} + 2 \approx 6.2$
3	$y = 3\sqrt{3} + 2 \approx 7.2$
4	$y = 3\sqrt{4} + 2 = 8$

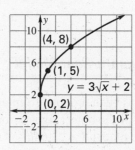

Exercises for Example 1

Find the domain and the range of the function. Then sketch the graph.

1. $y = 2\sqrt{x} + 1$ **2.** $y = 2\sqrt{x} - 1$ **3.** $y = 2\sqrt{x} - 2$

EXAMPLE 2 *Graphing $y = \sqrt{x - h}$*

Find the domain and the range of $y = \sqrt{x - 2}$. Then sketch its graph.

SOLUTION

To find the domain, find the values of x for which the radicand is nonnegative.

$x - 2 \geq 0$ Write an inequality for the domain.

$x \geq 2$ Add two to each side.

NAME _____ DATE _____

Reteaching with Practice

For use with pages 709–714

The domain is the set of all numbers that are greater than or equal to 2. The range is the set of all nonnegative numbers. Make a table of values, plot the points, and connect them with a smooth curve.

x	y
2	$y = \sqrt{2} - 2 = 0$
3	$y = \sqrt{3} - 2 = 1$
4	$y = \sqrt{4} - 2 \approx 1.4$
5	$y = \sqrt{5} - 2 \approx 1.7$
6	$y = \sqrt{6} - 2 = 2$

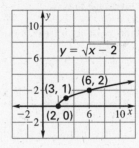

Exercises for Example 2

Find the domain and the range of the function. Then sketch its graph.

4. $y = \sqrt{x} - 1$ **5.** $y = \sqrt{x} + 1$ **6.** $y = \sqrt{x} - 4$

EXAMPLE 3   *Using a Square-Root Model*

An object has been dropped from a height of h feet. The speed S (in ft/sec) of the object right before it strikes the ground is given by the model $S = \sqrt{64h}$.

a. Sketch the graph of the model.

b. Find the speed S (in ft/sec) of an object that has been dropped from a height of 25 feet.

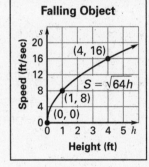

SOLUTION

a. Make a table of values, plot the points, and connect them with a smooth curve.

h	0	1	2	3	4
S	$\sqrt{64 \cdot 0} = 0$	$\sqrt{64 \cdot 1} = 8$	$\sqrt{64 \cdot 2} \approx 11.3$	$\sqrt{64 \cdot 3} \approx 13.9$	$\sqrt{64 \cdot 4} = 16$

b. Substitute $h = 25$ into the model: $S = \sqrt{64 \cdot 25} = 40$ ft/sec

Exercise for Example 3

7. Use the model in Example 3 to find the speed S (in ft/sec) of an object that has been dropped from a height of 36 feet.

NAME _____ DATE _____

Quick Catch-Up for Absent Students

For use with pages 709–715

The items checked below were covered in class on (date missed) _____

Lesson 12.1: Square-Root Functions

_____ **Goal 1:** Evaluate and graph a square-root function. (pp. 709–710)

Material Covered:

     _____ Activity: Investigating a Square-Root Function

     _____ Example 1: Graphing $y = a\sqrt{x}$

     _____ Example 2: Graphing $y = \sqrt{x} + k$

     _____ Student Help: Study Tip

     _____ Example 3: Graphing $y = \sqrt{x - h}$

Vocabulary:

     square-root function, p. 709

_____ **Goal 2:** Use square-root functions to model real-life problems. (p. 711)

Material Covered:

     _____ Example 4: Using a Square-Root Model

Activity 12.1: Families of Square-Root Functions (p. 715)

_____ **Goal:** Graph square-root functions using a graphing calculator or a computer.

     _____ Student Help: Keystroke Help

_____ Other (specify) _____

     _____

Homework and Additional Learning Support

     _____ Textbook (specify) _pp. 712–714_ _____

     _____

     _____ Internet: Extra Examples at www.mcdougallittell.com

     _____ *Reteaching with Practice* worksheet (specify exercises)_____

     _____ *Personal Student Tutor* for Lesson 12.1

NAME _____ DATE _____

Interdisciplinary Application

For use with pages 709–714

Spiral Shells

BIOLOGY Spirals occur in many forms in nature. Some are left-handed (spiraling counterclockwise) and some are right-handed (spiraling clockwise). Some snails occur in both left- and right-handed versions.

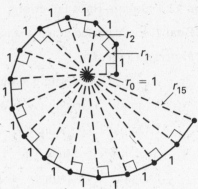

The square root spiral shown at the right is formed by a sequence of right triangles, each with a side whose length is 1. Let r_n be the length of the hypotenuse of the nth triangle.

Using the Pythagorean Theorem, $a^2 + b^2 = c^2$ where a and b are the legs of the triangle and c is the hypotenuse, you can find the length of $r_1, r_2, \ldots r_{15}$.

1. Substitute r for c in the Pythagorean Theorem.

2. Solve the equation in Exercise 1 for r.

3. The legs of the first triangle each have a length of 1. Find r_1.

4. For the second triangle, the lengths of the legs are $a = 1$ and $b =$ the value of r_1. Use these lengths to find r_2.

5. Copy and complete the table. Describe the pattern.

n	0	1	2	3	4	5	6	7	8	9	10	11	12	13	14	15
r_n																

6. Let A_n be the area of the nth triangle. Find a formula for A_n.

7. Find A_8. Give your answer in simplified form.

NAME _____ DATE _____

Challenge: Skills and Applications

For use with pages 709–714

In Exercises 1–6, find the domain of the function.

Example: $y = \sqrt{\dfrac{1}{x - 7}}$

Solution: To find the domain, find all values of x for which the radicand is nonnegative. There is no value of x that makes the radicand zero, so solve:

$$\frac{1}{x - 7} > 0$$

Since 1 is positive, the radicand is positive when the denominator is $x - 7 > 0 \Rightarrow x > 7$. The domain is the set of all numbers $x > 7$.

1. $y = \sqrt{\dfrac{1}{x + 9}}$

2. $y = \sqrt{\dfrac{-16}{3x + 8}}$

Example: $y = \sqrt{4 - x^2}$

Solution: To find the domain, find all values of x for which the radicand is nonnegative. $4 - x^2 \geq 0 \Rightarrow (2 - x)(2 + x) \geq 0$. The x-intercepts of the graph are at $x = 2$ and $x = -2$. The function is zero at these points and the function changes sign there. Therefore, it is necessary to test points in the intervals $x < -2$, $-2 < x < 2$, and $x > 2$. When $x = -3$, $2 - x$ is positive, $2 + x$ is negative, and so the product is negative. Testing $x = 0$ and $x = 3$ similarly, we find that $4 - x^2 \geq 0$ when $-2 \leq x \leq 2$. The domain is the set of all numbers in this interval.

3. $y = \sqrt{x^2 - 5x - 14}$

4. $y = \sqrt{x^2 - 9x}$

5. $y = \sqrt{6x - x^2}$

6. $y = \sqrt{-x^2 + 8x - 15}$

7. If the price of an item increases from p_1 to p_2 over a period of 2 years, the rate of inflation i for the item can be modeled by $i = \sqrt{\dfrac{p_2}{p_1}} - 1$. If a gallon of milk costs \$1.09 in 1996 and \$1.29 in 1998, what was the rate of inflation for milk?

TEACHER'S NAME _____ CLASS _____ ROOM _____ DATE _____

Lesson Plan

2-day lesson (See *Pacing the Chapter,* TE pages 706C–706D)   For use with pages 716–721

GOALS   1. **Add, subtract, multiply, and divide radical expressions.**
2. **Use radical expressions in real-life situations.**

State/Local Objectives _____

_____

✓ **Check the items you wish to use for this lesson.**

STARTING OPTIONS
____ Homework Check: TE page 712; Answer Transparencies
____ Warm-Up or Daily Homework Quiz: TE pages 716 and 714, CRB page 24, or Transparencies

TEACHING OPTIONS
____ Motivating the Lesson: TE page 717
____ Lesson Opener (Activity): CRB page 25 or Transparencies
____ Graphing Calculator Activity with Keystrokes: CRB pages 26
____ Examples: Day 1: 1–4, SE pages 716–717; Day 2: 5, SE page 718
____ Extra Examples: Day 1: TE page 717 or Transp.; Day 2: TE page 718 or Transp.; Internet
____ Closure Question: TE page 718
____ Guided Practice: SE page 719; Day 1: Exs. 1–16; Day 2: Ex. 17

APPLY/HOMEWORK
Homework Assignment
____ Basic   Day 1: 18–58 even; Day 2: 19–57 odd, 63–65, 70, 73, 76, 79
____ Average   Day 1: 18–58 even; Day 2: 19–57 odd, 60–65, 70, 73, 76, 79
____ Advanced   Day 1: 18–58 even; Day 2: 19–57 odd, 60–68, 70, 73, 76, 79

Reteaching the Lesson
____ Practice Masters: CRB pages 27–29 (Level A, Level B, Level C)
____ Reteaching with Practice: CRB pages 30–31 or Practice Workbook with Examples
____ Personal Student Tutor

Extending the Lesson
____ Applications (Real-Life): CRB page 33
____ Challenge: SE page 721; CRB page 34 or Internet

ASSESSMENT OPTIONS
____ Checkpoint Exercises: Day 1: TE page 717 or Transp.; Day 2: TE page 718 or Transp.
____ Daily Homework Quiz (12.2): TE page 721, CRB page 37, or Transparencies
____ Standardized Test Practice: SE page 721; TE page 721; STP Workbook; Transparencies

Notes _____

_____

_____

TEACHER'S NAME _____ CLASS _____ ROOM _____ DATE _____

Lesson Plan for Block Scheduling

1-day lesson (See *Pacing the Chapter,* TE pages 706C–706D)   For use with pages 716–721

GOALS   1. Add, subtract, multiply, and divide radical expressions.
2. Use radical expressions in real-life situations.

State/Local Objectives _____

_____

_____

✓ **Check the items you wish to use for this lesson.**

STARTING OPTIONS
____ Homework Check: TE page 712; Answer Transparencies
____ Warm-Up or Daily Homework Quiz: TE pages 716 and
         714, CRB page 24, or Transparencies

TEACHING OPTIONS
____ Motivating the Lesson: TE page 717
____ Lesson Opener (Activity): CRB page 25 or Transparencies
____ Graphing Calculator Activity with Keystrokes: CRB page 26
____ Examples: Day 1: 1–4, SE pages 716–717; Day 2: 5, SE page 718
____ Extra Examples: Day 1: TE page 717 or Transp.; Day 2: TE page 718 or Transp.; Internet
____ Closure Question: TE page 718
____ Guided Practice: SE page 719; Day 1: Exs. 1–16; Day 2: Ex. 17

APPLY/HOMEWORK
Homework Assignment (See also the assignments for Lessons 12.1 and 12.3.)
____ Block Schedule: Day 1: 18–58 even; Day 2: 19–57 odd, 60–65, 70, 73, 76, 79

Reteaching the Lesson
____ Practice Masters: CRB pages 27–29 (Level A, Level B, Level C)
____ Reteaching with Practice: CRB pages 30–31 or Practice Workbook with Examples
____ Personal Student Tutor

Extending the Lesson
____ Applications (Real-Life): CRB page 33
____ Challenge: SE page 721; CRB page 34 or Internet

ASSESSMENT OPTIONS
____ Checkpoint Exercises: Day 1: TE page 717 or Transp.; Day 2: TE page 718 or Transp.
____ Daily Homework Quiz (12.2): TE page 721, CRB page 37, or Transparencies
____ Standardized Test Practice: SE page 721; TE page 721; STP Workbook; Transparencies

Notes _____

_____

_____

CHAPTER PACING GUIDE	
Day	**Lesson**
1	12.1 (all); **12.2 (begin)**
2	**12.2 (end)**; 12.3 (begin)
3	12.3 (end); 12.4 (begin)
4	12.4 (end); 12.5 (begin)
5	12.5 (end); 12.6 (all)
6	12.7 (all); 12.8 (all)
7	Review/Assess Ch. 12

Lesson 12.2

NAME _____ DATE _____

WARM-UP EXERCISES

For use before Lesson 12.2, pages 716–721

1. Simplify $2a + a - 6a$.

2. Multiply $(x + 3)(x - 3)$.

3. Multiply $(2x - 9)(2x + 9)$.

4. Solve $x^2 - 5x + 3 = 0$.

....

DAILY HOMEWORK QUIZ

For use after Lesson 12.1, pages 709–715

1. Evaluate the square-root function for the given value of x. Round your answer to the nearest tenth.

   a. $\sqrt{x - 6}$; 20

   b. $9\sqrt{16 - x}$; −2

2. Find the domain and range of the function.

   a. $y = \sqrt{2x - 7}$

   b. $y = 7 + \sqrt{x}$

   c. $y = \dfrac{\sqrt{3 - x}}{x}$

Algebra 1
Chapter 12 Resource Book

NAME _____ DATE _____

Activity Lesson Opener

For use with pages 716–721

Fill in the blanks.

1. You know that $3x + 2x =$ ___.
   Let $x = \sqrt{2}$.
   Then $3\sqrt{2} + 2\sqrt{2} =$ ___.

2. You know that $4y + 3y =$ ___.
   Let $y = \sqrt{3}$.
   Then $4\sqrt{3} + 3\sqrt{3} =$ ___.

3. You know that $5x - x =$ ___.
   Let $x = \sqrt{5}$.
   Then $5\sqrt{5} - \sqrt{5} =$ ___.

4. You know that $8z - 6z + 4z =$ ___.
   Let $z = \sqrt{10}$.
   Then $8\sqrt{10} - 6\sqrt{10} + 4\sqrt{10} =$ ___.

5. You know that $3a + 5a - 2a =$ ___.
   Let $a = \sqrt{7}$.
   Then $3\sqrt{7} + 5\sqrt{7} - 2\sqrt{7} =$ ___.

6. You know that $x(x) =$ ___.
   Let $x = \sqrt{2}$.
   Then $\sqrt{2}(\sqrt{2}) =$ ___.

7. You know that $y(2y) =$ ___.
   Let $y = \sqrt{5}$.
   Then $\sqrt{5}(2\sqrt{5}) =$ ___.

8. You know that $6z(4z) =$ ___.
   Let $z = \sqrt{3}$.
   Then $6\sqrt{3}(4\sqrt{3}) =$ ___.

NAME _____ DATE _____

Graphing Calculator Activity Keystrokes

For use with page 720

Keystrokes for Exercise 62

TI-82

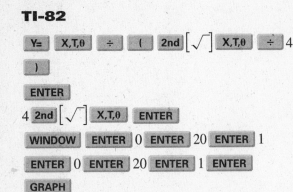

TI-83

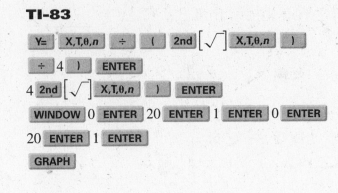

SHARP EL-9600c

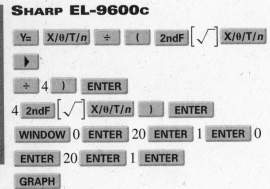

CASIO CFX-9850GA PLUS

From the main menu, choose GRAPH.

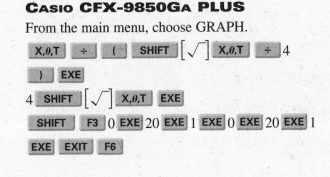

Algebra 1
Chapter 12 Resource Book

Practice A

For use with pages 716–721

Simplify the radical.

1. $\sqrt{32}$

2. $\sqrt{75}$

3. $\sqrt{27}$

4. $\sqrt{72}$

5. $\sqrt{20}$

6. $\sqrt{12}$

7. $\sqrt{125}$

8. $\sqrt{48}$

Simplify the expression.

9. $4\sqrt{3} - 2\sqrt{3}$

10. $2\sqrt{5} + \sqrt{5}$

11. $7\sqrt{5} - 3\sqrt{5}$

12. $8\sqrt{3} - \sqrt{3}$

13. $12\sqrt{6} + 5\sqrt{6} - 2\sqrt{6}$

14. $5\sqrt{8} - 6\sqrt{8}$

15. $\sqrt{32} + 2\sqrt{2}$

16. $\sqrt{12} - 2\sqrt{3}$

17. $\sqrt{18} - \sqrt{2}$

18. $\sqrt{20} + \sqrt{80}$

19. $\sqrt{28} - \sqrt{63}$

20. $\sqrt{24} + \sqrt{54} + 8\sqrt{6}$

Simplify the expression.

21. $\sqrt{4} \cdot \sqrt{4}$

22. $\sqrt{9} \cdot \sqrt{4}$

23. $\sqrt{8} \cdot \sqrt{4}$

24. $\sqrt{12} \cdot \sqrt{6}$

25. $\sqrt{5}(3\sqrt{2} + \sqrt{4})$

26. $\sqrt{3}(\sqrt{12} - 6\sqrt{3})$

27. $(1 - \sqrt{7})(1 + \sqrt{7})$

28. $(\sqrt{5} + 2)^2$

29. $\dfrac{3}{\sqrt{6}}$

30. $\dfrac{3}{\sqrt{2}}$

31. $\dfrac{7}{\sqrt{3}}$

32. $\dfrac{4}{2 + \sqrt{3}}$

Show whether the expression is a solution of the equation.

33. $x^2 + 2x - 5 = 0;\ -1 + \sqrt{6}$

34. $x^2 + 2x - 1 = 0;\ -1 + \sqrt{2}$

35. $x^2 - 4x + 1 = 0;\ 2 + \sqrt{3}$

36. $x^2 - 6x + 2 = 0;\ 3 + \sqrt{7}$

37. *Pole–vaulting*   A pole–vaulter's approach velocity v (in feet per second) and height reached h (in feet) are related by the following equation.

   Pole–vaulting model: $v = 8\sqrt{h}$

   Approximate how fast you were running if you vaulted 16 feet.

Geometry   **Find the area.**

38.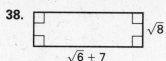

39.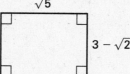

40.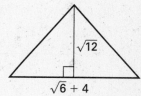

Lesson 12.2

Practice B

For use with pages 716–721

Simplify the radical.

1. $\sqrt{98}$ 2. $\sqrt{48}$ 3. $\sqrt{54}$ 4. $\sqrt{162}$

5. $\sqrt{192}$ 6. $\sqrt{112}$ 7. $\sqrt{250}$ 8. $\sqrt{108}$

Simplify the expression.

9. $4\sqrt{3} + 2\sqrt{3}$ 10. $8\sqrt{5} + \sqrt{5}$ 11. $7\sqrt{2} - 3\sqrt{2}$

12. $10\sqrt{6} - 13\sqrt{6}$ 13. $\sqrt{20} + \sqrt{5}$ 14. $\sqrt{48} - 3\sqrt{12}$

15. $\sqrt{18} + \sqrt{32}$ 16. $\sqrt{12} - \sqrt{48} + \sqrt{3}$ 17. $2\sqrt{8} - \sqrt{98} + \sqrt{72}$

18. $\sqrt{28} - 3\sqrt{7} + \sqrt{63}$ 19. $\sqrt{200} - \sqrt{242} - \sqrt{2}$ 20. $\sqrt{40} + \sqrt{90} - \sqrt{1000}$

Simplify the expression.

21. $\sqrt{8} \cdot \sqrt{2}$ 22. $\sqrt{12} \cdot \sqrt{3}$ 23. $\sqrt{9} \cdot \sqrt{7}$

24. $\sqrt{2}(3\sqrt{2} + \sqrt{5})$ 25. $\sqrt{7}(3\sqrt{5} - \sqrt{20})$ 26. $\sqrt{6}(2\sqrt{3} - 4\sqrt{2})$

27. $(\sqrt{3} + 2)^2$ 28. $(4 - \sqrt{5})^2$ 29. $(3\sqrt{2} - 1)^2$

30. $\dfrac{6}{\sqrt{3}}$ 31. $\dfrac{5}{\sqrt{5}}$ 32. $\dfrac{7}{\sqrt{2}}$

33. $\dfrac{3}{4 + \sqrt{2}}$ 34. $\dfrac{1}{\sqrt{3} - 2}$ 35. $\dfrac{2}{5 + \sqrt{2}}$

Show whether the expression is a solution of the equation.

36. $x^2 + 6x + 2 = 0; -3 + \sqrt{7}$ 37. $2x^2 - 6x + 3 = 0; 3 + \sqrt{3}$

38. $x^2 - 4x - 6 = 0; 1 - \sqrt{10}$ 39. $x^2 - 8x + 3 = 0; 4 - \sqrt{13}$

40. *Science Center*   A new science center opens. For the first 10 weeks, the number of people that visit the center can be modeled by

$$N = \sqrt{5000 + 320t}$$

where N is in hundreds of people and $t = 0$ corresponds to week 1. How many people visited the center the opening week? Make a table that shows the number of visitors over the first 10 weeks.

41. *Geometry*   Find the area and the perimeter of the rectangle.

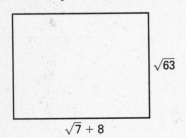

$\sqrt{63}$

$\sqrt{7} + 8$

NAME _____ DATE _____

Practice C

For use with pages 716–721

Simplify the expression.

1. $\sqrt{3} - 5\sqrt{3}$

2. $10\sqrt{3} + 5\sqrt{3}$

3. $6\sqrt{6} + 5\sqrt{6}$

4. $\sqrt{8} - \sqrt{32}$

5. $7\sqrt{45} + 2\sqrt{20}$

6. $\sqrt{\frac{1}{2}} - \sqrt{\frac{1}{8}}$

7. $2\sqrt{3} + 3\sqrt{12}$

8. $2\sqrt{10} - 3\sqrt{40} + 4\sqrt{5}$

9. $\sqrt{10} + 3\sqrt{10} - \sqrt{5}$

10. $\sqrt{128} - 3\sqrt{2} + \sqrt{32}$

11. $2\sqrt{50} - \sqrt{8} + 3\sqrt{18}$

12. $3\sqrt{27} + 5\sqrt{48} - \sqrt{300}$

13. $2\sqrt{20} - 3\sqrt{24} - \sqrt{180}$

14. $\sqrt{500} - \sqrt{245} + \sqrt{320}$

15. $\sqrt{160} + \sqrt{250} + \sqrt{3600}$

Simplify the expression.

16. $\sqrt{7} \cdot \sqrt{12}$

17. $\sqrt{32} \cdot \sqrt{5}$

18. $\sqrt{32} \cdot \sqrt{12}$

19. $\sqrt{2}(\sqrt{18} + 4\sqrt{3})$

20. $\sqrt{5}(7\sqrt{5} - \sqrt{10})$

21. $2\sqrt{3}(2\sqrt{3} - 3\sqrt{2})$

22. $(\sqrt{3} + 2)(\sqrt{3} - 2)$

23. $(4 - \sqrt{5})(4 + \sqrt{5})$

24. $(5\sqrt{3} - 2)^2$

25. $\dfrac{4}{\sqrt{7}}$

26. $\dfrac{15}{2\sqrt{6}}$

27. $\dfrac{-24}{\sqrt{6}}$

28. $\dfrac{2}{8 + \sqrt{11}}$

29. $\dfrac{3}{2\sqrt{5} - 3}$

30. $\dfrac{3}{6 + \sqrt{5}}$

Show whether the expression is a solution of the equation.

31. $x^2 - 4x + 1 = 0; 2 - \sqrt{3}$

32. $x^2 + 10x + 5 = 0; -5 + \sqrt{5}$

33. $x^2 + 5x - 1 = 0; -5 - \sqrt{29}$

34. $x^2 + 10x - 2 = 0; -5 - 3\sqrt{3}$

Geometry. Find the area and perimeter of each figure.

35.

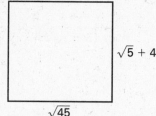

$\sqrt{5} + 4$

$\sqrt{45}$

36.

3   $\sqrt{12}$

$\sqrt{3}$

37.

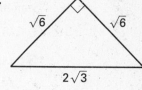

$\sqrt{6}$ $\sqrt{6}$

$2\sqrt{3}$

38. *Electricity* The voltage V required for a circuit is given by $V = \sqrt{PR}$ where P is the power in watts and R is the resistance in ohms. Find the volts needed to light a 75-watt bulb with a resistance of 110 ohms.

NAME _____ DATE _____

Reteaching with Practice

For use with pages 716–721

GOAL Add, subtract, multiply, and divide radical expressions and use radical expressions in real-life situations

---

VOCABULARY

The product $(a + \sqrt{b})(a - \sqrt{b})$ does not involve radicals. The expressions $(a + \sqrt{b})$ and $(a - \sqrt{b})$ are **conjugates** of each other.

EXAMPLE 1 *Adding and Subtracting Radicals*

Simplify the expression.

$\sqrt{12} + \sqrt{3}$.

SOLUTION

$$\sqrt{12} + \sqrt{3} = \sqrt{4 \cdot 3} + \sqrt{3} \qquad \text{Perfect square factor}$$
$$= \sqrt{4} \cdot \sqrt{3} + \sqrt{3} \qquad \text{Use product property.}$$
$$= 2\sqrt{3} + \sqrt{3} \qquad \text{Simplify.}$$
$$= 3\sqrt{3} \qquad \text{Add like radicals.}$$

Exercises for Example 1

Simplify the expression.

1. $\sqrt{7} + 3\sqrt{7}$ 2. $\sqrt{8} - \sqrt{2}$ 3. $\sqrt{48} + \sqrt{3}$

EXAMPLE 2 *Multiplying Radicals*

Simplify the expression.

a. $\sqrt{3} \cdot \sqrt{12}$ **b.** $\sqrt{5}(\sqrt{2} + \sqrt{3})$ **c.** $(3 + \sqrt{2})(3 - \sqrt{2})$

SOLUTION

a. $\sqrt{3} \cdot \sqrt{12} = \sqrt{36}$ Use product property.
$\qquad\qquad = 6$ Simplify.

b. $\sqrt{5}(\sqrt{2} + \sqrt{3}) = \sqrt{5} \cdot \sqrt{2} + \sqrt{5} \cdot \sqrt{3}$ Use distributive property.
$\qquad\qquad\qquad = \sqrt{10} + \sqrt{15}$ Use product property.

c. $(3 + \sqrt{2})(3 - \sqrt{2}) = 3^2 - (\sqrt{2})^2$ Use sum and difference pattern.
$\qquad\qquad\qquad = 9 - 2 = 7$ Simplify.

Exercises for Example 2

Simplify the expression.

4. $(\sqrt{2} + 1)^2$ 5. $\sqrt{3} \cdot \sqrt{6}$ 6. $\sqrt{10}(2 + \sqrt{2})$

Reteaching with Practice

For use with pages 716–721

EXAMPLE 3 *Simplifying Radicals*

Simplify $\dfrac{5}{\sqrt{2}}$.

SOLUTION

$$\dfrac{5}{\sqrt{2}} = \dfrac{5}{\sqrt{2}} \cdot \dfrac{\sqrt{2}}{\sqrt{2}} \qquad \text{Multiply numerator and denominator by } \sqrt{2}.$$

$$= \dfrac{5\sqrt{2}}{\sqrt{2} \cdot \sqrt{2}} \qquad \text{Multiply fractions.}$$

$$= \dfrac{5\sqrt{2}}{2} \qquad \text{Simplify.}$$

Exercises for Example 3

Simplify the expression.

7. $\dfrac{4}{\sqrt{3}}$

8. $\dfrac{5}{\sqrt{8}}$

9. $\dfrac{-1}{\sqrt{12}}$

EXAMPLE 4 *Using a Radical Model*

A tsunami is an enormous ocean wave that can be caused by underwater earthquakes, volcanic eruptions, or hurricanes. The speed S of a tsunami in miles per hour is given by the model $S = 3.86\sqrt{d}$ where d is the depth of the ocean in feet. Suppose one tsunami is at a depth of 1792 feet and another is at a depth of 1372 feet. Write an expression that represents the difference in speed between the tsunamis. Simplify the expression.

SOLUTION

The speed of the first tsunami mentioned is $3.86\sqrt{1792}$ while the speed of the second tsunami is $3.86\sqrt{1372}$. The difference D between the speeds can be represented by $3.86\sqrt{1792} - 3.86\sqrt{1372}$.

$$D = 3.86\sqrt{1792} - 3.86\sqrt{1372}$$

$$= 3.86\sqrt{7 \cdot 256} - 3.86\sqrt{7 \cdot 196}$$

$$= 61.76\sqrt{7} - 54.04\sqrt{7} = 7.72\sqrt{7}$$

Exercise for Example 4

10. Rework Example 4 if one tsunami is at a depth of 3125 feet and another tsunami is at a depth of 2000 feet.

Quick Catch-Up for Absent Students

For use with pages 716–721

The items checked below were covered in class on (date missed) _____

Lesson 12.2: Operations with Radical Expressions

_____ **Goal 1:** Add, subtract, multiply, and divide radical expressions. (pp. 716–717)

Material Covered:

_____ Example 1: Adding and Subtracting Radicals

_____ Example 2: Multiplying Radicals

_____ Example 3: Simplifying Radicals

_____ Example 4: Checking Quadratic Formula Solutions

Vocabulary:

like radicals, p. 716   conjugates, p. 717

_____ **Goal 2:** Use radical expressions in real-life situations. (p. 718)

Material Covered:

_____ Example 5: Using a Radical Model

_____ Other (specify) _____

_____

Homework and Additional Learning Support

_____ Textbook (specify) _pp. 719–721_____

_____

_____ Internet: Extra Examples at www.mcdougallittell.com

_____ *Reteaching with Practice* worksheet (specify exercises)_____

_____ *Personal Student Tutor* for Lesson 12.2

NAME _____ DATE _____

Real-Life Application:
When Will I Ever Use This?

For use with pages 716–721

Plywood

Plywood is a building material usually made of an odd number of thin layers of wood glued together. The layers, called plies, are arranged so that the grain direction (direction of the wood fibers) of each layer are at right angles to that of the next layer. The outside plies are called faces and backs. The center ply or plies are called the core.

Plywood's main advantages over ordinary lumber are that it is lightweight and workable, yet rigid and strong. Plywood can be cut to exact sizes. Plywood is produced in large panels for ease of application, strength, and smooth surfaces. It shrinks and swells less than ordinary wood and has a greater resistance to splitting at the ends. Plywood is used chiefly as a structure upon which finished walls, flooring, and roofing are laid.

In Exercises 1-4, use the following information.

Plywood is sold in a large rectangular shape. The length of the plywood is $\left(\sqrt{39} + 2\right)$ feet and the width is $\sqrt{18}$ feet.

1. Draw a diagram showing the length and width of a piece of plywood.

2. Find the area in square feet.

3. Find the number of pieces of plywood needed to cover an area of 315 square feet.

4. Find the number of rectangular pieces measuring $\sqrt{5}$ ft $\times$ $\sqrt{8}$ ft that can be cut from one piece of plywood.

In Exercises 5 and 6, use the diagram below.

5. The drawing to the right shows a floor plan with the plywood already in place. Find the perimeter.

6. Find the area of the floor plan.

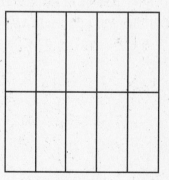

Algebra 1
Chapter 12 Resource Book

In Exercises 1–7, simplify the expression.

1. $\dfrac{5}{\sqrt{3}} + \dfrac{7}{\sqrt{12}}$

2. $\dfrac{2\sqrt{3}}{\sqrt{15}} - \dfrac{4}{\sqrt{20}}$

3. $3\sqrt{2}\left(5x\sqrt{7} - 4\sqrt{2}\right)$

4. $-2x\sqrt{3}\left(4\sqrt{12} + \sqrt{3}\right)$

5. $\dfrac{2}{\sqrt{5}}\left(8\sqrt{30} - \sqrt{120}\right)$

6. $7x\sqrt{5}\left(\sqrt{20} - 3\sqrt{5}\right)$

7. $\left(x\sqrt{3} - \sqrt{6}\right)\left(2x\sqrt{3} + \sqrt{6}\right)$

8. Suppose a quantity called i has the property that, in any calcualtion, i^2 could always be replaced by -1, but otherwise i behaves like any variable. Simplify each expression. Give your answer in terms of i.

   a. $(3 + 4i)^2$

   b. $(5 - 8i)(7 + 4i)$

   c. $(2 + 5i)(2 - 5i)$

   d. Generalize part (c) by expressing $(a + bi)(a - bi)$ in terms of a and b.

In Exercises 9–14, use the figure. Write answers as a radical expression in simplest form.

9. Find the perimeter of triangle *BCD*.

10. Find the perimeter of triangle *ABD*.

11. How much longer is the perimeter of triangle *BCD* than the perimeter of triangle *ABD*?

12. Find the area of triangle *BCD*.

13. Find the area of triangle *ABD*.

14. If the area of triangle *ABD* is $2\sqrt{6}$, what is x?

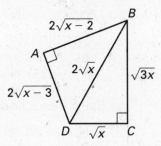

TEACHER'S NAME _____ CLASS _____ ROOM _____ DATE _____

Lesson Plan

2-day lesson (See *Pacing the Chapter,* TE pages 706C–706D)   For use with pages 722–728

GOALS   1. **Solve a radical equation.**
2. **Use radical equations to solve real-life problems.**

State/Local Objectives _____

_____

✓ **Check the items you wish to use for this lesson.**

STARTING OPTIONS
____ Homework Check: TE page 719; Answer Transparencies
____ Warm-Up or Daily Homework Quiz: TE pages 722 and 721, CRB page 37, or Transparencies

TEACHING OPTIONS
____ Motivating the Lesson: TE page 723
____ Lesson Opener (Activity): CRB page 38 or Transparencies
____ Graphing Calculator Activity with Keystrokes: CRB pages 39–40
____ Examples: Day 1: 1–4, SE pages 722–723; Day 2: 5, SE page 724
____ Extra Examples: Day 1: TE page 723 or Transp.; Day 2: TE page 724 or Transp.
____ Closure Question: TE page 724
____ Guided Practice: SE page 725; Day 1: Exs. 1–13; Day 2: Ex. 14

APPLY/HOMEWORK
Homework Assignment
____ Basic   Day 1: 16–60 even, 66, 74, 76; Day 2: 15–61 odd, 67, 69, 75, 77–80, 85, 90, 95, 99, 100;
          Quiz 1: 1–16
____ Average   Day 1: 16–60 even, 66, 74, 76; Day 2: 15–61 odd, 67, 69, 75, 77–80, 85, 90, 95, 99, 100;
          Quiz 1: 1–16
____ Advanced   Day 1: 16–60 even, 66, 74, 76; Day 2: 15–61 odd, 67, 69, 75, 77–81, 85, 90, 95, 99,
          100; Quiz 1: 1–16

Reteaching the Lesson
____ Practice Masters: CRB pages 41–43 (Level A, Level B, Level C)
____ Reteaching with Practice: CRB pages 44–45 or Practice Workbook with Examples
____ Personal Student Tutor

Extending the Lesson
____ Applications (Interdisciplinary): CRB page 47
____ Challenge: SE page 727; CRB page 48 or Internet

ASSESSMENT OPTIONS
____ Checkpoint Exercises: Day 1: TE page 723 or Transp.; Day 2: TE page 724 or Transp.
____ Daily Homework Quiz (12.3): TE page 727, CRB page 52, or Transparencies
____ Standardized Test Practice: SE page 727; TE page 727; STP Workbook; Transparencies
____ Quiz (12.1–12.3): SE page 728; CRB page 49

Notes _____

_____

_____

TEACHER'S NAME _____ CLASS _____ ROOM _____ DATE _____

Lesson Plan for Block Scheduling

1-day lesson (See *Pacing the Chapter,* TE pages 706C–706D)   For use with pages 722–728

GOALS   1. Solve a radical equation.
2. Use radical equations to solve real-life problems.

State/Local Objectives _____

_____

_____

✓ **Check the items you wish to use for this lesson.**

STARTING OPTIONS
_____ Homework Check: TE page 719; Answer Transparencies
_____ Warm-Up or Daily Homework Quiz: TE pages 722 and
         721, CRB page 37, or Transparencies

TEACHING OPTIONS
_____ Motivating the Lesson: TE page 723
_____ Lesson Opener (Activity): CRB page 38 or Transparencies
_____ Graphing Calculator Activity with Keystrokes: CRB pages 39–40
_____ Examples: Day 2: 1–4, SE pages 722–723; Day 3: 5, SE page 724
_____ Extra Examples: Day 2: TE page 723 or Transp.; Day 3: TE page 724 or Transp.
_____ Closure Question: TE page 724
_____ Guided Practice: SE page 725; Day 2: Exs. 1–13; Day 3: Ex. 14

APPLY/HOMEWORK
Homework Assignment (See also the assignments for Lessons 12.2 and 12.4.)
_____ Block Schedule: Day 2: 16–60 even, 66, 74, 76;
                 Day 3: 15–61 odd, 67, 69, 75, 77–80, 85, 90, 95, 99, 100; Quiz 1: 1–16

Reteaching the Lesson
_____ Practice Masters: CRB pages 41–43 (Level A, Level B, Level C)
_____ Reteaching with Practice: CRB pages 44–45 or Practice Workbook with Examples
_____ Personal Student Tutor

Extending the Lesson
_____ Applications (Interdisciplinary): CRB page 47
_____ Challenge: SE page 727; CRB page 48 or Internet

ASSESSMENT OPTIONS
_____ Checkpoint Exercises: xDay 2: TE page 723 or Transp.; Day 3: TE page 724 or Transp.
_____ Daily Homework Quiz (12.3): TE page 727, CRB page 52, or Transparencies
_____ Standardized Test Practice: SE page 727; TE page 727; STP Workbook; Transparencies
_____ Quiz (12.1–12.3): SE page 728; CRB page 49

Notes _____

_____

_____

CHAPTER PACING GUIDE	
Day	**Lesson**
1	12.1 (all); 12.2 (begin)
2	12.2 (end); **12.3 (begin)**
3	**12.3 (end)**; 12.4 (begin)
4	12.4 (end); 12.5 (begin)
5	12.5 (end); 12.6 (all)
6	12.7 (all); 12.8 (all)
7	Review/Assess Ch. 12

LESSON 12.3

NAME _____ DATE _____

WARM-UP EXERCISES

For use before Lesson 12.3, pages 722–728

1. Simplify $\left(\sqrt{6}\right)^2$.

2. Solve $\sqrt{x} = 4$.

3. Solve $\sqrt{x} = 17$.

4. Solve $x^2 - 5x + 6 = 0$.

5. Solve $\dfrac{x}{3} = \dfrac{3}{5}$.

DAILY HOMEWORK QUIZ

For use after Lesson 12.2, pages 716–721

1. Simplify the expression.

   a. $5\sqrt{6} - 9\sqrt{6}$

   b. $3\sqrt{17} + 9\sqrt{11} + \sqrt{17}$

   c. $\sqrt{7} \cdot \sqrt{3}$

   d. $\sqrt{2}\left(7\sqrt{3} + \sqrt{2}\right)$

   e. $\left(\sqrt{3} + 4\right)^2$

   f. $\dfrac{5}{8 - \sqrt{6}}$

2. Solve $x^2 - 8x - 1 = 0$.

Algebra 1
Chapter 12 Resource Book

37

Lesson 12.3

Activity Lesson Opener

For use with pages 722–728

SET UP: Work with a partner.

Complete the solution.

1. The square root of a number is 4. Find the number.

$$\sqrt{x} = 4$$
$$(\sqrt{x})^2 = 4^2$$
$$\underline{\hspace{2cm}} = \underline{\hspace{1.5cm}}$$

2. The square root of a number is 10. Find the number.

$$\sqrt{x} = 10$$
$$(\sqrt{x})^2 = 10^2$$
$$\underline{\hspace{2cm}} = \underline{\hspace{1.5cm}}$$

3. $\quad \sqrt{x} = 7$

$$(\sqrt{x})^2 = 7^2$$
$$\underline{\hspace{2cm}} = \underline{\hspace{1.5cm}}$$

4. $\quad \sqrt{x} = 6$

$$(\sqrt{x})^2 = \underline{\hspace{1cm}}^2$$
$$\underline{\hspace{2cm}} = \underline{\hspace{1.5cm}}$$

5. $\quad \sqrt{x} = 5$

$$\underline{\hspace{1.5cm}}^2 = 5^2$$
$$\underline{\hspace{2cm}} = \underline{\hspace{1.5cm}}$$

6. $\quad \sqrt{x} = 3$

$$\underline{\hspace{1.5cm}}^2 = \underline{\hspace{1cm}}^2$$
$$\underline{\hspace{2cm}} = \underline{\hspace{1.5cm}}$$

7. Make a conjecture about how to solve an equation of the form $\sqrt{a} = b$.

Graphing Calculator Activity Keystrokes

For use with page 726

Keystrokes for Exercise 56

TI-82

[Y=] [2nd] [√] [(] [X,T,θ] [+] [4] [)]
[ENTER]
3 [ENTER]
[ZOOM] 6
[2nd] [CALC] 5 [ENTER] [ENTER]
[ENTER]

TI-83

[Y=] [2nd] [√] [X,T,θ,n] [+] [4] [)] [ENTER]
3 [ENTER]
[ZOOM] 6
[2nd] [CALC] 5 [ENTER] [ENTER]
[ENTER]

SHARP EL-9600c

[Y=] [2ndF] [√] [X/θ/T/n] [+] 4 [ENTER]
3 [ENTER]
[ZOOM] [A] 5
[2ndF] [CALC] 2

CASIO CFX-9850Ga PLUS

From the main menu, choose GRAPH.

[X,θ,T] [SHIFT] [√] [(] [X,θ,T] [+] [4] [)] [EXE]
3 [EXE]
[SHIFT] [F3] [F3] [EXIT]
[F6] [F5] [F5]

Algebra 1
Chapter 12 Resource Book

39

Graphing Calculator Activity Keystrokes

For use with page 726

Keystrokes for Exercise 65

TI-82

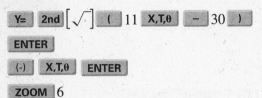

ENTER

(-) X,T,θ ENTER

ZOOM 6

TI-83

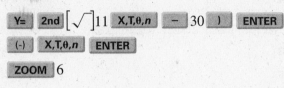

(-) X,T,θ,n ENTER

ZOOM 6

SHARP EL-9600c

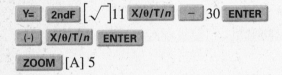

(-) X/θ/T/n ENTER

ZOOM [A] 5

CASIO CFX-9850GA PLUS

From the main menu, choose GRAPH.

SHIFT [√] (X,θ,T − 30) EXE

(-) X,θ,T EXE

SHIFT F3 F3 EXIT

F6

NAME _____ DATE _____

Practice A

For use with pages 722–728

Isolate the radical expression on one side of the equation. Do not solve.

1. $\sqrt{x+2} - 4 = 8$

2. $\sqrt{x} - 7 = 9$

3. $\sqrt{x-7} - 2 = 0$

4. $\sqrt{5x-4} - 3 = -1$

5. $-\sqrt{2x-1} = 0$

6. $-8 + \sqrt{x} = -4$

7. $\sqrt{\frac{1}{2}x - 4} - 5 = 7$

8. $\sqrt{4-x} + \frac{3}{4} = \frac{5}{4}$

9. $2\sqrt{x} - 4 = 8$

Solve the equation. Check for extraneous solutions.

10. $\sqrt{x} - 5 = 0$

11. $\sqrt{x} - 7 = 0$

12. $\sqrt{x} + 3 = 12$

13. $\sqrt{x} - 14 = -8$

14. $\sqrt{x} + 6 = 0$

15. $\sqrt{2x} = 0$

16. $\sqrt{x+3} = 2$

17. $\sqrt{2x-5} = 13$

18. $\sqrt{x-2} = 8$

19. $\sqrt{2x+1} = 0$

20. $\sqrt{8x+1} = 5$

21. $\sqrt{4x+1} - 3 = 0$

22. $\sqrt{3x-5} + 7 = 3$

23. $\sqrt{x+2} = x - 4$

24. $\sqrt{5x^2 - 7} = 2x$

25. $4 + \sqrt{x-2} = x$

26. $\sqrt{1-2x} = 1 + x$

27. $x = \sqrt{5x-4}$

Two numbers and their geometric mean are given. Find the value of *a*.

28. 8 and *a*; 12

29. 7 and *a*; 14

30. 5 and *a*; 15

31. *Water Flow*   You can measure the speed of water using an L-shaped tube. The speed *V* of the water in miles per hour is $V = \sqrt{2.5h}$ where *h* is the height of the column of water above the surface in inches. If you use the tube in a river and find that *h* is 5 inches, what is the speed of the water to the nearest tenth of a mile per hour?

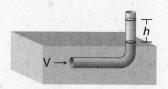

32. *Cylindrical Rod*   The radius *r*, the height *h*, and the surface area *A* of a cylinder are related by the equation

$$r = \sqrt{\frac{A}{2\pi} - rh}.$$

What is the radius of a rod whose surface area is 50π square inches and height is 24 inches?

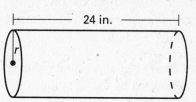

33. Find the value of *x* if the perimeter is 24.

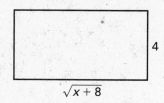

Algebra 1
Chapter 12 Resource Book

41

Lesson 12.3

NAME _____ DATE _____

Practice B

For use with pages 722–728

Solve the equation. Check for extraneous solutions.

1. $\sqrt{x} - 5 = 0$ **2.** $\sqrt{x} + 6 = 0$ **3.** $\sqrt{5x} - 2 = 3$

4. $\sqrt{3x + 4} + 1 = 11$ **5.** $\sqrt{9 - x} - 4 = 2$ **6.** $\sqrt{2x + 1} + 4 = 3$

7. $\sqrt{\dfrac{1}{3}x + 2} = 8$ **8.** $\sqrt{\dfrac{3}{2}x - 5} - 10 = -8$ **9.** $5 - \sqrt{4x - 3} = 3$

10. $\sqrt{\dfrac{x}{6}} = 2$ **11.** $\sqrt{\dfrac{4x}{5}} - 9 = 3$ **12.** $5\sqrt{\dfrac{4x}{3}} - 2 = 0$

13. $4 - \sqrt{2x + 3} = 5$ **14.** $\sqrt{\dfrac{2}{5}x + 3} - \dfrac{4}{5} = \dfrac{2}{5}$ **15.** $6 + \sqrt{7x + 3} = 6$

16. $x = \sqrt{12 - x}$ **17.** $x = \sqrt{2x + 3}$ **18.** $x = \sqrt{8 - 2x}$

19. $x = \sqrt{-3x - 2}$ **20.** $x = \sqrt{24 - 5x}$ **21.** $x = \sqrt{6x - 8}$

22. $x = \sqrt{9x - 14}$ **23.** $2x = \sqrt{16x - 15}$ **24.** $2x = \sqrt{-8x - 4}$

25. $4x = \sqrt{2x + 5}$ **26.** $\dfrac{1}{2}x = \sqrt{8 - x}$ **27.** $\dfrac{1}{2}x = \sqrt{2x - 4}$

Two numbers and their geometric mean are given. Find the value of *a*.

28. 8 and a; 16   **29.** 3 and a; 15   **30.** 9 and a; 6

31. *Market Research* A marketing department determines that the price of a magazine subscription and the demand to subscribe are related by the equation

$$P = 52 - \sqrt{0.0002x + 1}$$

where P is the price per subscription and x is the number of subscriptions sold. If the subscription price is set at $30, how many subscriptions would be sold?

32. *Free Falling Velocity* The velocity v of a free falling object, the height h in which it falls, and the acceleration due to gravity g (32 feet per second squared) are related by the equation $v = \sqrt{2gh}$. Find the height from which a penny was dropped if it strikes the ground with a velocity of 30 feet per second.

33. *Oriental Fan* The area A and radius r of the fully opened fan shown are related by the equation

$$r = \sqrt{A}.$$

If the radius of the fan is 12 inches, find the area of the fully opened fan.

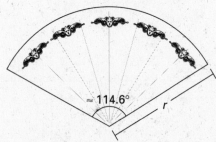

$\approx 114.6°$

34. If the radius of the fan in Exercise 33 is 6 inches, is the area half the area of the fan with radius 12 inches? Explain.

Solve the equation. Check for extraneous solutions.

1. $\sqrt{x} - 10 = 0$

2. $\sqrt{x} - 12 = 0$

3. $\sqrt{-x} - \dfrac{1}{2} = \dfrac{3}{2}$

4. $\sqrt{3x - 2} = 5$

5. $\sqrt{3x + 9} - 12 = 0$

6. $\sqrt{4x - 3} - 5 = 0$

7. $\sqrt{3x} - 4 = 6$

8. $\sqrt{3x + 2} + 2 = 3$

9. $-5 + \sqrt{4 - x} = 1$

10. $4 = 6 - \sqrt{21x - 3}$

11. $10 = 17 + \sqrt{6x + 7}$

12. $\sqrt{\dfrac{1}{2}x - 5} - 1 = 11$

13. $\sqrt{\dfrac{1}{9}x + 1} - \dfrac{2}{3} = \dfrac{5}{3}$

14. $-5 - \sqrt{10x - 2} = 5$

15. $6 - \sqrt{7x - 9} = 3$

16. $x = \sqrt{20 - x}$

17. $x = \sqrt{6 - x}$

18. $2x = \sqrt{-10x - 4}$

19. $x = \sqrt{100 - 15x}$

20. $x = \sqrt{77 - 4x}$

21. $2x = \sqrt{-13x - 10}$

22. $\dfrac{1}{2}x = \sqrt{x + 3}$

23. $1 + x = \sqrt{1 - 2x}$

24. $x = 4 + \sqrt{x - 2}$

25. $x - 2 = \sqrt{2x - 1}$

26. $3x = \sqrt{3 - 6x}$

27. $2x = \sqrt{4x + 15}$

Use a graphing calculator to graphically solve the radical equation. Check the solution algebraically.

28. $x + 3 = \sqrt{x^2 + 5}$

29. $x - 2 = \sqrt{x^2 - 4}$

30. $x + 2 = \sqrt{8x + 1}$

31. $6 = -4\sqrt{2x - 1} - 10$

32. $2\sqrt{x + 5} + 5 = 7$

33. $3\sqrt{x - 4} + 5 = 11$

The Speed of Sound   **In Exercises 34 and 35, use the following information.**

The speed of sound near Earth's surface depends on the temperature. An equation that relates the speed v (in meters per second) with the temperature t (in degrees Celsius) is $v = 20\sqrt{t + 273}$.

34. Your friend is playing basketball 170 meters from you. You hear the sound of the ball hitting the backboard 0.5 seconds after seeing the ball hit the backboard. What is the temperature?

35. The temperature $-273°C$ is called absolute zero. What is the speed at this temperature?

36. *Pendulumn*   The time it takes for one swing of a pendulum is

$$T = 2\pi\sqrt{\dfrac{L}{32}}$$

where T is the time in seconds and L is the length of the pendulum in feet. Find the length of a pendulum that makes one swing in 3 seconds. Round to the nearest hundredth.

NAME _____ DATE _____

Reteaching with Practice

For use with pages 722–728

GOAL Solve a radical equation and use radical equations to solve real-life problems

EXAMPLE 1 *Solving a Radical Equation*

Solve $\sqrt{3x + 1} + 2 = 6$.

SOLUTION

Isolate the radical expression on one side of the equation.

$\sqrt{3x + 1} + 2 = 6$	Write original equation.
$\sqrt{3x + 1} = 4$	Subtract 2 from each side.
$\left(\sqrt{3x + 1}\right)^2 = 4^2$	Square each side.
$3x + 1 = 16$	Simplify.
$3x = 15$	Subtract 1 from each side.
$x = 5$	Divide each side by 3.

The solution is 5.

Exercises for Example 1

Solve the equation.

1. $\sqrt{x + 2} = 3$ **2.** $\sqrt{x} + 2 = 3$ **3.** $\sqrt{4x + 1} = 3$

EXAMPLE 2 *Checking for Extraneous Solutions*

Solve the equation $\sqrt{2x + 3} = x$.

SOLUTION

$\sqrt{2x + 3} = x$	Write original equation.
$\left(\sqrt{2x + 3}\right)^2 = x^2$	Square each side.
$2x + 3 = x^2$	Simplify.
$0 = x^2 - 2x - 3$	Write in standard form.
$0 = (x - 3)(x + 1)$	Factor.
$x = 3 \text{ or } x = -1$	Zero-product property.

To check the solutions, substitute 3 and -1 in the original equation.

$$\sqrt{2(3) + 3} \stackrel{?}{=} 3 \qquad\qquad \sqrt{2(-1) + 3} \stackrel{?}{=} -1$$
$$\sqrt{9} \stackrel{?}{=} 3 \qquad\qquad\qquad \sqrt{1} \stackrel{?}{=} -1$$
$$3 = 3 \qquad\qquad\qquad\quad 1 \neq -1$$

The only solution is 3, because $x = -1$ is not a solution.

Reteaching with Practice

For use with pages 722–728

Exercises for Example 2

Solve the equation and check for extraneous solutions.

4. $\sqrt{x-1} + 3 = x$ **5.** $\sqrt{3x} + 6 = 0$ **6.** $\sqrt{x+6} = x$

EXAMPLE 3 ## Using a Radical Model

The distance d (in centimeters) that tap water is absorbed up a strip of blotting paper at a temperature of 28.4°C is given by the model

$$d = 0.444\sqrt{t} \text{ where } t \text{ is the time (in seconds).}$$

Approximately how many minutes would it take for the water to travel a distance of 16 centimeters up the strip of blotting paper?

SOLUTION

$d = 0.444\sqrt{t}$	Write model for blotting paper distance.
$16 = 0.444\sqrt{t}$	Substitute 16 for d.
$\dfrac{16}{0.444} = \sqrt{t}$	Divide each side by 0.444.
$\left(\dfrac{16}{0.444}\right)^2 = t$	Square each side.
$1299 \approx t$	Use a calculator.

It would take approximately 1299 seconds for the water to travel a distance of 16 centimeters up the strip of blotting paper. To find the time in minutes, you divide 1299 by 60. It would take approximately 22 minutes.

Exercises for Example 3

7. Use the model in Example 3 to find the distance that the water would travel in 36 seconds.

8. Use the model in Example 3 to find the number of seconds that it would take for the water to travel a distance of 10 centimeters up the strip of blotting paper.

NAME _____ DATE _____

Quick Catch-Up for Absent Students

For use with pages 722–728

The items checked below were covered in class on (date missed) _____

Lesson 12.3: Solving Radical Equations

_____ **Goal 1:** Solve a radical equation. (pp. 722–723)

Material Covered:

_____ Example 1: Solving a Radical Equation

_____ Example 2: Solving a Radical Equation

_____ Student Help: Study Tip

_____ Example 3: Checking for Extraneous Solutions

_____ Example 4: Using a Geometric Mean

_____ **Goal 2:** Use radical equations to solve real-life problems. (p. 724)

Material Covered:

_____ Example 5: Using a Radical Model

_____ Other (specify) _____

_____

Homework and Additional Learning Support

_____ Textbook (specify) _pp. 725–728_____

_____

_____ *Reteaching with Practice* worksheet (specify exercises)_____

_____ *Personal Student Tutor* for Lesson 12.3

NAME _____ DATE _____

Interdisciplinary Application

For use with pages 722–728

Balance Scales

CHEMISTRY   Balance scales can give inaccurate results if the lengths of the left and right arms, L and R, are not *exactly* the same. The following is a technique that scientists use to be sure of an accurate weight. Let W be the true weight of an object.

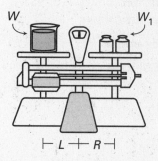

1. Place the object on the left side of the scale. Counterbalance the right side with a weight of W_1.

   By a property of levers, you can write $\dfrac{W}{W_1} = \dfrac{R}{L}$.

2. Place the object on the right side of the scale. Counterbalance the left side with a weight of W_2.

   By a property of levers, you can write $\dfrac{W_2}{W} = \dfrac{R}{L}$.

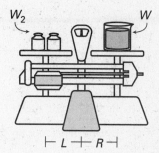

3. Equate the two expressions for $\dfrac{R}{L}$ and solve for W.

   The solution is $W = \sqrt{W_1 W_2}$. Thus, the true weight is the geometric mean of the weights obtained by weighing the object on the left and right sides of the scale.

In Exercises 1-3, use the information above.

1. In science class, you weigh a sulfur sample only once on the left side of a scale. It weighs 92 grams. Your teacher says that the actual weight is 95 grams. If 95 grams is correct, then what weight would you have obtained on the right side of the scale?

2. You weigh a mercury sample only once on the right side of a scale. It weighs 6 ounces. You know that the actual weight of the mercury sample is 5.7 ounces. Find the weight you would have obtained on the left side of the scale.

3. You are given a sample of a substance. On the right side of the scale, the substance weighs $x + 1$ grams. On the left side of the scale, the substance weighs $x - 2$ grams. Find the measurements given by your balance scale if $W^2 = 1054$ grams.

NAME _____ DATE _____

Challenge: Skills and Applications

For use with pages 722–728

In Exercises 1–6, solve the equation. Check the extraneous solutions.

Example: $\sqrt{x} - 7 = \sqrt{x - 7}$

Solution: $x - 14\sqrt{x} + 49 = x - 7$

$-14\sqrt{x} = -56 \Rightarrow \sqrt{x} = 4 \Rightarrow x = 16$

Check: $\sqrt{16} - 7 = \sqrt{16 - 7}$

$4 - 7 = \sqrt{9}$

$-3 \neq 3$

There is no solution.

1. $\sqrt{x + 5} = \sqrt{2x - 13}$

2. $\sqrt{x^2 + 7x - 9} = \sqrt{x^2 + 4x + 12}$

3. $\sqrt{x} + 4 = \sqrt{x + 4}$

4. $\sqrt{x} - 3 = \sqrt{x + 3}$

5. $\sqrt{x} - 1 = \sqrt{x - 7}$

6. $\sqrt{x} - 4 = \sqrt{x - 8}$

7. If the geometric mean of x and $2x$ is 6, what is the value of x?

8. If the geometric mean of x and $3x$ is 6, what is the value of x?

9. If the geometric mean of x and $4x$ is 6, what is the value of x?

10. Given that the geometric mean of x and ax is 6 and $a > 0$, as a increases does the absolute value is x increase or decrease?

In Exercises 11 and 12, use the following information.

Melissa Winston and George Gilsen are designing rectangular boxes with bases that are square. They have 17.1 square meters of cardboard to use for the faces of each box. The length of a side of the square base of a box s can be modeled by $s = -h + \sqrt{h^2 + 8.55}$, where h is the height of the box in meters.

11. Find the length of a side of a box's base when the height is 1.5 meters.

12. Find the height of a box when the length of a base side is 1.5 meters.

NAME _____ DATE _____

Quiz 1

For use after Lessons 12.1–12.3

1. Evaluate $y = \frac{1}{3}\sqrt{x} - 2$ when $x = 9$. *(Lesson 12.1)*

2. Sketch the graph of $y = \sqrt{x} + 3$. *(Lesson 12.1)*

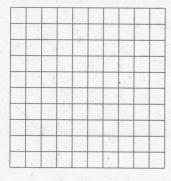

In Exercises 3 and 4, simplify the expression. *(Lesson 12.2)*

3. $2\sqrt{3} + \sqrt{12}$

4. $\left(3 + \sqrt{5}\right)\left(3 - \sqrt{5}\right)$

5. Solve the equation. *(Lesson 12.3)*

$$\sqrt{6x - 3} + 1 = 10$$

6. If the geometric mean of 5 and a is 15, what is the value of a?
   (Lesson 12.3)

Answers

1. _____

2. Use grid at left.

3. _____

4. _____

5. _____

6. _____

Lesson 12.3

Algebra 1
Chapter 12 Resource Book

TEACHER'S NAME _____ CLASS _____ ROOM _____ DATE _____

Lesson Plan

2-day lesson (See *Pacing the Chapter,* TE pages 706C–706D)   **For use with pages 729–736**

GOALS   **1. Solve a quadratic equation by completing the square.**
           2. Choose a method for solving a quadratic equation.

State/Local Objectives _____

_____

✓ **Check the items you wish to use for this lesson.**

STARTING OPTIONS
_____ Homework Check: TE page 725; Answer Transparencies
_____ Warm-Up or Daily Homework Quiz: TE pages 730 and 727, CRB page 52, or Transparencies

TEACHING OPTIONS
_____ Motivating the Lesson: TE page 731
_____ Concept Activity: SE page 729; CRB page 53 (Activity Support Master)
_____ Lesson Opener (Application): CRB page 54 or Transparencies
_____ Examples: Day 1: 1–4, SE pages 730–731; Day 2: 5–6, SE page 733
_____ Extra Examples: Day 1: TE page 731 or Transp.; Day 2: TE page 732 or Transp.; Internet
_____ Closure Question: TE page 733
_____ Guided Practice: SE page 734; Day 1: Exs. 1–13; Day 2: Exs. 14–19

APPLY/HOMEWORK
Homework Assignment
_____ Basic   Day 1: 20–66 even; Day 2: 33–67 odd, 68–70, 74–76, 82, 86, 90, 94, 98, 102, 106
_____ Average   Day 1: 20–66 even; Day 2: 33–67 odd, 68–71, 74–76, 82, 86, 90, 94, 98, 102, 106
_____ Advanced   Day 1: 20–66 even; Day 2: 33–67 odd, 68–80, 82, 86, 90, 94, 98, 102, 106

Reteaching the Lesson
_____ Practice Masters: CRB pages 55–57 (Level A, Level B, Level C)
_____ Reteaching with Practice: CRB pages 58–59 or Practice Workbook with Examples
_____ Personal Student Tutor

Extending the Lesson
_____ Applications (Real-Life): CRB page 61
_____ Challenge: SE page 736; CRB page 62 or Internet

ASSESSMENT OPTIONS
_____ Checkpoint Exercises: Day 1: TE page 731 or Transp.; Day 2: TE page 732 or Transp.
_____ Daily Homework Quiz (12.4): TE page 736, CRB page 65, or Transparencies
_____ Standardized Test Practice: SE page 736; TE page 736; STP Workbook; Transparencies

Notes _____

_____

_____

Algebra 1
Chapter 12 Resource Book

TEACHER'S NAME _____ CLASS _____ ROOM _____ DATE _____

Lesson Plan for Block Scheduling

1-day lesson (See *Pacing the Chapter,* TE pages 706C–706D)   For use with pages 729–736

GOALS
1. Solve a quadratic equation by completing the square.
2. Choose a method for solving a quadratic equation.

State/Local Objectives _____

_____

_____

✓ **Check the items you wish to use for this lesson.**

STARTING OPTIONS

_____ Homework Check: TE page 725; Answer Transparencies
_____ Warm-Up or Daily Homework Quiz: TE pages 730 and
      727, CRB page 52, or Transparencies

TEACHING OPTIONS

_____ Motivating the Lesson: TE page 731
_____ Concept Activity: SE page 729; CRB page 53 (Activity Support Master)
_____ Lesson Opener (Application): CRB page 54 or Transparencies
_____ Examples: Day 3: 1–4, SE pages 730–731; Day 4: 5–6, SE page 733
_____ Extra Examples: Day 3: TE page 731 or Transp.; Day 4: TE page 732 or Transp.; Internet
_____ Closure Question: TE page 733
_____ Guided Practice: SE page 734; Day 3: Exs. 1–13; Day 4: Exs. 14–19

APPLY/HOMEWORK

Homework Assignment (See also the assignments for Lessons 12.3 and 12.5.)
_____ Block Schedule: Day 3: 20–66 even; Day 4: 33–67 odd, 68–71, 74–76, 82, 86, 90, 94, 98, 102, 106

Reteaching the Lesson
_____ Practice Masters: CRB pages 55–57 (Level A, Level B, Level C)
_____ Reteaching with Practice: CRB pages 58–59 or Practice Workbook with Examples
_____ Personal Student Tutor

Extending the Lesson
_____ Applications (Real-Life): CRB page 61
_____ Challenge: SE page 736; CRB page 62 or Internet

ASSESSMENT OPTIONS

_____ Checkpoint Exercises: Day 3: TE page 731 or Transp.; Day 4: TE page 732 or Transp.
_____ Daily Homework Quiz (12.4): TE page 736, CRB page 65, or Transparencies
_____ Standardized Test Practice: SE page 736; TE page 736; STP Workbook; Transparencies

Notes _____

_____

_____

CHAPTER PACING GUIDE	
Day	Lesson
1	12.1 (all); 12.2 (begin)
2	12.2 (end); 12.3 (begin)
3	12.3 (end); **12.4 (begin)**
4	**12.4 (end)**; 12.5 (begin)
5	12.5 (end); 12.6 (all)
6	12.7 (all); 12.8 (all)
7	Review/Assess Ch. 12

NAME _____ DATE _____

WARM-UP EXERCISES

For use before Lesson 12.4, pages 729–736

Simplify.

1. $(x + 9)^2$

2. $(2x - 3)^2$

Factor.

3. $x^2 - 10x + 25$

4. $9x^2 + 24x + 16$

DAILY HOMEWORK QUIZ

For use after Lesson 12.3, pages 722–728

1. Solve the equation.

   a. $\sqrt{x} - 5 = 0$

   b. $\sqrt{2x - 7} - 5 = 5$

   c. $x = \sqrt{15x - 14}$

2. The geometric mean of 3 and a is 9. What is a?

3. The perimeter of a rectangle is 20. Its dimensions are 4 and $\sqrt{x - 1}$. Find the value of x.

NAME _____ DATE _____

Activity Support Master

For use with page 729

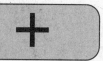
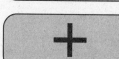
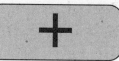

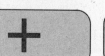
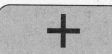
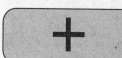

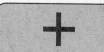

Application Lesson Opener

For use with pages 730–736

A rectangular pool is 2 ft longer than it is wide. The bottom of the pool has an area of 35 ft². If *x* represents the width of the pool, *x*(*x* + 2) = 35 models the area.

1. Simplify the left side of the equation.

2. Add the square of one-half of the coefficient of x to both sides.

3. Factor the trinomial on the left side of the equation. How could you use this information to solve the equation?

A rectangular window has a width that is 4 in. less than its height. The area of the window is 480 in². If *x* represents the height of the window, *x*(*x* − 4) = 480 models the area.

4. Simplify the left side of the equation.

5. Add the square of one-half of the coefficient of x to both sides.

6. Factor the trinomial on the left side of the equation. How could you use this information to solve the equation?

The length of the floor of a rectangular room is 6 ft greater than its width. The floor has an area of 216 ft². If *x* represents the width of the floor, *x*(*x* + 6) = 216 models the area.

7. Simplify the left side of the equation.

8. Add the square of one-half of the coefficient of x to both sides.

9. Factor the trinomial on the left side of the equation. How could you use this information to solve the equation?

Algebra 1
Chapter 12 Resource Book

Practice A

For use with pages 730–736

Identify the leading coefficient, the coefficient of *x*, and the constant term.

1. $x^2 + 6x + 9$

2. $x^2 - 8x + 16$

3. $x^2 - 12x + 36$

4. $25 + 10x + x^2$

5. $9 + 12x + 4x^2$

6. $1 - 10x + 25x^2$

Find the term that should be added to the expression to create a perfect square trinomial.

7. $x^2 + 4x$

8. $x^2 + 14x$

9. $x^2 - 8x$

10. $x^2 - 2x$

11. $x^2 + 10x$

12. $x^2 - 16x$

13. $x^2 + 7x$

14. $x^2 - 3x$

15. $x^2 + \frac{1}{2}x$

16. $x^2 - 0.8x$

17. $x^2 - 20x$

18. $x^2 + \frac{2}{3}x$

Solve the equation by completing the square.

19. $x^2 + 6x = 7$

20. $x^2 + 4x = 5$

21. $x^2 - 8x = 9$

22. $x^2 - 12x + 20 = 0$

23. $x^2 - 4x + 2 = 0$

24. $x^2 + 4x - 1 = 0$

25. $x^2 + 6x - 4 = 0$

26. $x^2 - 2x - 5 = 0$

27. $x^2 + 3x - 2 = 0$

28. $x^2 + x - 1 = 0$

29. $x^2 + \frac{1}{2}x - 2 = 0$

30. $2x^2 + 6x - 2 = 0$

Solve the quadratic equation.

31. $3x^2 - 12 = 0$

32. $x^2 + 9x + 20 = 0$

33. $6x^2 + 12x = 0$

34. $x^2 + x - 5 = 0$

35. $3x^2 - 2x - 8 = 0$

36. $3x^2 + 2x - 10 = 0$

Geometry **In Exercises 37 and 38, find the dimensions of the figure.**

37.

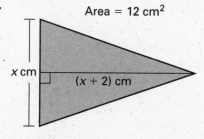

Area = 12 cm²

x cm   (*x* + 2) cm

38.

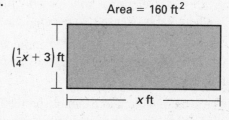

Area = 160 ft²

$\left(\frac{1}{4}x + 3\right)$ ft

x ft

39. *Flight of an Arrow* An arrow is shot into the air with an upward velocity of 48 feet per second from a hill 32 feet high. The height *h* of the arrow can be found using the model $h = 32 + 48t - 16t^2$ where *t* is the time in seconds. How many seconds after release will the arrow be 64 feet above the ground?

Practice B

For use with pages 730–736

Write the trinomial as the square of a binomial.

1. $x^2 - 2x + 1$

2. $x^2 + 24x + 144$

3. $x^2 + 14x + 49$

4. $x^2 + \frac{4}{3}x + \frac{4}{9}$

5. $x^2 - x + \frac{1}{4}$

6. $x^2 - 12x + 36$

Find the term that should be added to the expression to create a perfect square trinomial.

7. $x^2 - 6x$

8. $x^2 + 8x$

9. $x^2 - 22x$

10. $x^2 - \frac{1}{2}x$

11. $x^2 + 18x$

12. $x^2 - 30x$

13. $x^2 + 9x$

14. $x^2 - 7x$

15. $x^2 + \frac{2}{5}x$

16. $x^2 - 1.6x$

17. $x^2 - 24x$

18. $x^2 + \frac{1}{3}x$

Solve the equation by completing the square.

19. $x^2 + 7x = -12$

20. $x^2 - 5x = -4$

21. $x^2 + 14x = 15$

22. $x^2 - 24x + 9 = 0$

23. $x^2 - 8x - 4 = 0$

24. $x^2 - 10x - 23 = 0$

25. $2x^2 - 5x + 1 = 0$

26. $5x^2 + 10x - 7 = 0$

27. $3x^2 - 7x - 3 = 0$

28. $6x^2 - 7x - 3 = 0$

29. $9x^2 - 6x - 2 = 0$

30. $4x^2 - 4x - 5 = 0$

Solve the quadratic equation.

31. $x^2 + 4x - 1 = 0$

32. $5x^2 + 20x = 0$

33. $4x^2 - 4 = 0$

34. $3x^2 - 6x - 3 = 0$

35. $x^2 + 3x - 2 = 0$

36. $2x^2 + 2x - 14 = 0$

37. $3x^2 + 3x - 3 = 0$

38. $6x^2 + 7x - 20 = 0$

39. $7x^2 + x - 3 = 0$

40. *Baseball*   After a baseball is hit, the height h (in feet) of the ball above the ground t seconds after it is hit can be approximated by the equation $h = -16t^2 + 70t + 4$. Determine how long it will take for the ball to hit the ground. Round your answer to two decimal places.

41. *Rectangle*   The length of a rectangle is 2 centimeters less than twice the width. The area of the rectangle is 180 square centimeters. Find the length and width of the rectangle.

42. *Colorado*   The state of Colorado is almost perfectly rectangular, with its north border 111 miles longer than its west border. If the state encompasses 104,000 square miles, estimate the dimensions of Colorado. Round to the nearest mile.

Practice C

For use with pages 730–736

Write the trinomial as the square of a binomial.

1. $x^2 - 6x + 9$ 　　　　2. $x^2 + 26x + 169$ 　　　　3. $x^2 - 18x + 81$

4. $x^2 + \frac{6}{5}x + \frac{9}{25}$ 　　　5. $x^2 + x + \frac{1}{4}$ 　　　　6. $49 - 28x + 4x^2$

Find the term that should be added to the expression to create a perfect square trinomial.

7. $x^2 - 16x$ 　　　　　8. $x^2 + 18x$ 　　　　　9. $x^2 - 13x$

10. $x^2 - \frac{3}{4}x$ 　　　　11. $x^2 + 11x$ 　　　　12. $x^2 - 40x$

13. $x^2 + \frac{2}{9}x$ 　　　　14. $x^2 - 15x$ 　　　　15. $x^2 + \frac{3}{5}x$

Solve the equation by completing the square.

16. $x^2 + 6x = -5$ 　　　17. $x^2 + 8x - 4 = 5$ 　　　18. $x^2 - 6x + 5 = -4$

19. $x^2 + 4x - 7 = 0$ 　　20. $x^2 + 6x - 1 = 0$ 　　　21. $x^2 - 6x + 7 = 0$

22. $x^2 + 8x + 13 = 0$ 　　23. $x^2 - 3x + 1 = 0$ 　　　24. $x^2 - x - 4 = 0$

25. $3x^2 - 6x - 1 = 0$ 　　26. $2x^2 - 4x - 5 = 0$ 　　27. $4x^2 + 6x - 3 = 4$

28. $5x^2 - 8x + 4 = 10$ 　　29. $6x^2 - 12x - 2 = 8$ 　　30. $12x^2 + 30x - 6 = 0$

Solve the quadratic equation.

31. $x^2 - 48 = 0$ 　　　　32. $7x^2 - 14x = -3$ 　　　33. $12x^2 - 25x + 12 = 0$

34. $3x^2 + 10x = 8$ 　　　35. $x^2 - 6x - 2 = 0$ 　　　36. $3x^2 - 6x - 8 = -6$

37. $0.3x^2 + 0.1x - 0.2 = 0$ 　38. $6x^2 - 15x - 16 = -13$ 　39. $7x^2 - 12x = 0$

Projectiles **In Exercises 40 and 41, use the following information.**

The height of a projectile fired upward is given by the formula $s = v_0t - 16t^2$, where s is the height in feet, v_0 is the initial velocity in feet per second, and t is the time in seconds.

40. If the initial velocity is 200 feet per second, how long will it take for the projectile to return to Earth?

41. If the initial velocity is 128 feet per second, how much time will elapse until the projectile reaches a height of 64 feet? Round to the nearest hundredth of a second.

42. *Stopping Distance* A car with good tire tread can stop in less distance than a car with poor tread. The formula for the stopping distance d (in feet) of a car with good tread on dry cement is approximated by $d = 0.04v^2 + 0.5v$, where v is the speed of the car in miles per hour. If the driver must be able to stop within 60 feet, what is the maximum safe speed of the car? Round to the nearest mile per hour.

NAME _____ DATE _____

Reteaching with Practice

For use with pages 730–736

GOAL **Solve a quadratic equation by completing the square and choose a method for solving a quadratic equation**

VOCABULARY

To complete the square of the expression $x^2 + bx$, add the square of half the coefficient of x.

$$x^2 + bx + \left(\frac{b}{2}\right)^2 = \left(x + \frac{b}{2}\right)^2$$

EXAMPLE 1 ## *The Leading Coefficient is Not 1*

Solve $9x^2 - 18x + 5 = 0$ by completing the square.

SOLUTION

When the quadratic equation's leading coefficient is not 1, divide each side of the equation by this coefficient *before* completing the square.

$9x^2 - 18x + 5 = 0$	Write original equation.
$9x^2 - 18x = -5$	Subtract 5 from each side.
$x^2 - 2x = -\dfrac{5}{9}$	Divide each side by 9.
$x^2 - 2x + \left(\dfrac{2}{2}\right)^2 = -\dfrac{5}{9} + 1$	Add $\left(\dfrac{2}{2}\right)^2$, or 1 to each side.
$(x - 1)^2 = \dfrac{4}{9}$	Write left side as perfect square.
$(x - 1) = \pm\dfrac{2}{3}$	Find square root of each side.
$x = 1 \pm \dfrac{2}{3}$	Add 1 to each side.

The solutions are $\dfrac{5}{3}$ and $\dfrac{1}{3}$. Both solutions check in the original equation.

Exercises for Example 1

Solve the equation by completing the square.

1. $2n^2 - 3n = 2$　　　　**2.** $3y^2 + 4y = -1$　　　　**3.** $4b^2 + 8b + 3 = 0$

Reteaching with Practice

For use with pages 730–736

EXAMPLE 2 *Choosing a Solution Method*

Choose a method to solve the quadratic equation.

a. $5x^2 + 3x - 2 = 0$ **b.** $x^2 + 6x - 1 = 0$

SOLUTION

a. This equation can be factored easily.

$5x^2 + 3x - 2 = 0$	Write original equation.
$(5x - 2)(x + 1) = 0$	Factor.
$5x - 2 = 0$ or $x + 1 = 0$	Zero-product property.
$x = \dfrac{2}{5}$ or $x = -1$	Solve for x.

The solutions are $\dfrac{2}{5}$ and -1.

b. When this equation is written as $ax^2 + bx + c = 0$, $a = 1$ and b is an even number. Therefore it can be solved by completing the square.

$x^2 + 6x - 1 = 0$	Write original equation.
$x^2 + 6x = 1$	Add 1 to each side.
$x^2 + 6x + \left(\dfrac{6}{2}\right)^2 = 1 + 9$	Add $\left(\dfrac{6}{2}\right)^2$, or 9 to each side.
$(x + 3)^2 = 10$	Write left side as perfect square.
$x + 3 = \pm\sqrt{10}$	Find square root of each side.
$x = -3 \pm \sqrt{10}$	Subtract 3 from each side.

The solutions are $-3 + \sqrt{10}$ and $-3 - \sqrt{10}$.

Exercises for Example 2

Choose a method to solve the quadratic equation. Explain your choice.

4. $5y^2 - 35 = 0$ **5.** $w^2 - 3w - 10 = 0$ **6.** $2.7x^2 + 0.5x - 7 = 0$

NAME _____ DATE _____

Quick Catch-Up for Absent Students

For use with pages 729–736

The items checked below were covered in class on (date missed) _____

Activity 12.4: Modeling Completing the Square (p. 729)

_____ **Goal:** Use algebra tiles to complete a perfect square trinomial.

     _____ Student Help: Look Back

Lesson 12.4: Completing the Square

_____ **Goal 1:** Solve a quadratic equation by completing the square. (pp. 730–731)

Material Covered:

     _____ Example 1: Completing the Square

     _____ Student Help: Study Tip

     _____ Example 2: Solving a Quadratic Equation

     _____ Student Help: Study Tip

     _____ Example 3: Solving a Quadratic Equation

     _____ Example 4: The Leading Coefficient is Not 1

Vocabulary:

     completing the square, p. 730

_____ **Goal 2:** Choose a method for solving a quadratic equation. (pp. 732–733)

Material Covered:

     _____ Activity: Investigating the Quadratic Formula

     _____ Example 5: Choosing a Solution Method

     _____ Example 6: Choosing the Quadratic Formula

_____ Other (specify) _____

_____

Homework and Additional Learning Support

     _____ Textbook (specify) _pp. 734–736_____

_____

     _____ Internet: Extra Examples at www.mcdougallittell.com

     _____ *Reteaching with Practice* worksheet (specify exercises)_____

     _____ *Personal Student Tutor* for Lesson 12.4

Algebra 1
Chapter 12 Resource Book

Real-Life Application:
When Will I Ever Use This?

For use with pages 730–736

Caves

A cave, also known as a cavern, is a naturally hollow area in the Earth that is large enough for a person to enter. Some caves consist of a single chamber only a few yards deep. Other caves are vast networks of passages and chambers.

The longest cave ever explored is the Mammoth-Flint Ridge cave system in Kentucky. This cave has about 340 miles of explored and mapped passageways, but geologists think that it extends even farther.

The interior of a cave is a dark, damp place where sunlight never enters. Many caves have underground lakes, rivers, and waterfalls. Some of the most spectacular caves are popular tourist attractions. These caves have been equipped with pathways and electric lights. However, thousands of caves remain in their natural state, and many new caves and passages are being discovered each year.

In Exercises 1-3, use the following information.

You are exploring a new cave. You are asked to write down as much information about the cave as you can. You find yourself in a cavern that is large enough to stand in. The cross section showing the ceiling of the cavern can be modeled by the equation $y = -x^2 + 8x$, where y is the height of the cavern (in feet) and x is the distance from the left side of the cavern (in feet).

1. Use completing the square to solve the equation $-x^2 + 8x = 0$.

2. Sketch a graph of the equation $y = -x^2 + 8x$ to check the solution to Exercise 1.

3. Choose another method (finding square roots, using the quadratic formula, or factoring) for solving quadratic equations to check the solutions to Exercise 1. Explain your choice.

In Exercises 4 and 5, use the following information.

You enter a chamber with many stalactites. Stalactites are iciclelike formations that hang from the ceiling of a cave. The cross section showing the longest stalactite at its lowest point can be modeled by the equation $y = 12x^2 - 72x + 96$, where y is the length of the stalactite (in feet) and x is the horizontal distance from the left side of the stalactite (in feet).

4. Use completing the square to solve the equation $12x^2 - 72x = -96$.

5. Choose another method (finding square roots, graphing, using the quadratic formula, or factoring) for solving quadratic equations to check the solutions to Exercise 4. Explain your choice.

Lesson 12.4

Challenge: Skills and Applications

For use with pages 730–736

1. Solve the equation $x^2 - 8x = c$ for x by completing the square.

2. For what values of c does the equation from Exercise 1 have no solution?

3. Solve the equation $3x^2 - 5x = 3c$ for x by completing the square.

4. For what values of c does the equation from Exercise 3 have no solution?

5. Solve the equation $ax^2 - 6x = 4$ for x by completing the square.

6. For what values of a does the equation from Exercise 5 have no solution?

7. Solve the equation $x^2 - bx = -7$ for x by completing the square.

8. For what values of b does the equation from Exercise 7 have no solution?

In Exercises 9–11, use the following information.

The path of a pole vaulter making a jump is given by

$$y = -16x^2 + 8x + 16,$$

where y is the height of the pole vaulter in feet x seconds after the jump starts.

9. Write the equation in the form $y = a(x - h)^2 + k$.

10. What is the value of h in the equation from Exercise 9 and what does it mean in this situation?

11. What is the value of k in the equation from Exercise 9 and what does it mean in this situation?

TEACHER'S NAME _____ CLASS _____ ROOM _____ DATE _____

Lesson Plan

2-day lesson (See *Pacing the Chapter,* TE pages 706C–706D)   For use with pages 737–744

GOALS 1. **Use the Pythagorean Theorem and its converse.**
         2. **Use the Pythagorean Theorem and its converse in real-life problems.**

State/Local Objectives _____

_____

✓ **Check the items you wish to use for this lesson.**

STARTING OPTIONS

_____ Homework Check: TE page 734; Answer Transparencies
_____ Warm-Up or Daily Homework Quiz: TE pages 738 and 736, CRB page 65, or Transparencies

TEACHING OPTIONS

_____ Motivating the Lesson: TE page 739
_____ Concept Activity: SE page 737; CRB page 66 (Activity Support Master)
_____ Lesson Opener (Visual Approach): CRB page 67 or Transparencies
_____ Graphing Calculator Activity with Keystrokes: CRB pages 68–69
_____ Examples: Day 1: 1–3, SE pages 738–739; Day 2: 4–5, SE page 740
_____ Extra Examples: Day 1: TE page 739 or Transp.; Day 2: TE page 740 or Transp.
_____ Closure Question: TE page 740
_____ Guided Practice: SE page 741; Day 1: Exs. 1–12; Day 2: none

APPLY/HOMEWORK

Homework Assignment

_____ Basic   Day 1: 13–27; Day 2: 33–39, 41–45, 50, 55, 60, 65, 70, 71; Quiz 2: 1–10
_____ Average   Day 1: 13–27; Day 2: 33–39, 41–45, 50, 55, 60, 65, 70, 71; Quiz 2: 1–10
_____ Advanced   Day 1: 13–27; Day 2: 33–48, 50, 55, 60, 65, 70, 71; Quiz 2: 1–10

Reteaching the Lesson

_____ Practice Masters: CRB pages 70–72 (Level A, Level B, Level C)
_____ Reteaching with Practice: CRB pages 73–74 or Practice Workbook with Examples
_____ Personal Student Tutor

Extending the Lesson

_____ Applications (Real-Life): CRB page 76
_____ Math & History: SE page 744; CRB page 77; Internet
_____ Challenge: SE page 743; CRB page 78 or Internet

ASSESSMENT OPTIONS

_____ Checkpoint Exercises: Day 1: TE page 739 or Transp.; Day 2: TE page 740 or Transp.
_____ Daily Homework Quiz (12.5): TE page 743, CRB page 82, or Transparencies
_____ Standardized Test Practice: SE page 743; TE page 743; STP Workbook; Transparencies
_____ Quiz (12.4–12.5): SE page 744; CRB page 79

Notes _____

_____

_____

TEACHER'S NAME _____ CLASS _____ ROOM _____ DATE _____

Lesson Plan for Block Scheduling

1-day lesson (See *Pacing the Chapter*, TE pages 706C–706D)   For use with pages 737–744

GOALS
1. **Use the Pythagorean Theorem and its converse.**
2. **Use the Pythagorean Theorem and its converse in real-life problems.**

State/Local Objectives _____

_____

_____

✓ Check the items you wish to use for this lesson.

STARTING OPTIONS
_____ Homework Check: TE page 734; Answer Transparencies
_____ Warm-Up or Daily Homework Quiz: TE pages 738 and
          736, CRB page 65, or Transparencies

TEACHING OPTIONS
_____ Motivating the Lesson: TE page 739
_____ Concept Activity: SE page 737; CRB page 66 (Activity Support Master)
_____ Lesson Opener (Visual Approach): CRB page 67 or Transparencies
_____ Graphing Calculator Activity with Keystrokes: CRB pages 68–69
_____ Examples: Day 4: 1–3, SE pages 738–739; Day 5: 4–5, SE page 740
_____ Extra Examples: Day 4: TE page 739 or Transp.; Day 5: TE page 740 or Transp.
_____ Closure Question: TE page 740
_____ Guided Practice: SE page 741; Day 4: Exs. 1–12; Day 5: none

APPLY/HOMEWORK
Homework Assignment (See also the assignments for Lessons 12.4 and 12.6.)
_____ Block Schedule: Day 4: 13–27; Day 5: 33–39, 41–45, 50, 55, 60, 65, 70, 71; Quiz 2: 1–10

Reteaching the Lesson
_____ Practice Masters: CRB pages 70–72 (Level A, Level B, Level C)
_____ Reteaching with Practice: CRB pages 73–74 or Practice Workbook with Examples
_____ Personal Student Tutor

Extending the Lesson
_____ Applications (Real-Life): CRB page 76
_____ Math & History: SE page 744; CRB page 77; Internet
_____ Challenge: SE page 743; CRB page 78 or Internet

ASSESSMENT OPTIONS
_____ Checkpoint Exercises: Day 4: TE page 739 or Transp.; Day 5: TE page 740 or Transp.
_____ Daily Homework Quiz (12.5): TE page 743, CRB page 82, or Transparencies
_____ Standardized Test Practice: SE page 743; TE page 743; STP Workbook; Transparencies
_____ Quiz (12.4–12.5): SE page 744; CRB page 79

Notes _____

_____

_____

CHAPTER PACING GUIDE	
Day	**Lesson**
1	12.1 (all); 12.2 (begin)
2	12.2 (end); 12.3 (begin)
3	12.3 (end); 12.4 (begin)
4	12.4 (end); **12.5 (begin)**
5	**12.5 (end)**; 12.6 (all)
6	12.7 (all); 12.8 (all)
7	Review/Assess Ch. 12

WARM-UP EXERCISES

For use before Lesson 12.5, pages 737–744

Give the positive solution of each equation. Round to the nearest tenth where necessary.

1. $x^2 = 225$

2. $x^2 = 1700$

3. $\sqrt{400} = x$

4. $\sqrt{185} = x$

DAILY HOMEWORK QUIZ

For use after Lesson 12.4 pages 729–736

1. What term should be added to $x^2 - 16x$ to make it a perfect square trinomial?

2. Solve $x^2 - 20x - 2 = 0$.

3. Solve $4x^2 - 2x - 1 = 0$.

4. The height of a triangle is 2 cm more than its base. The area of the triangle is 30 cm^2. What are the base and height of the triangle?

LESSON

12.5

Lesson 12.5

NAME _____ DATE _____

Activity Support Master

For use with page 737

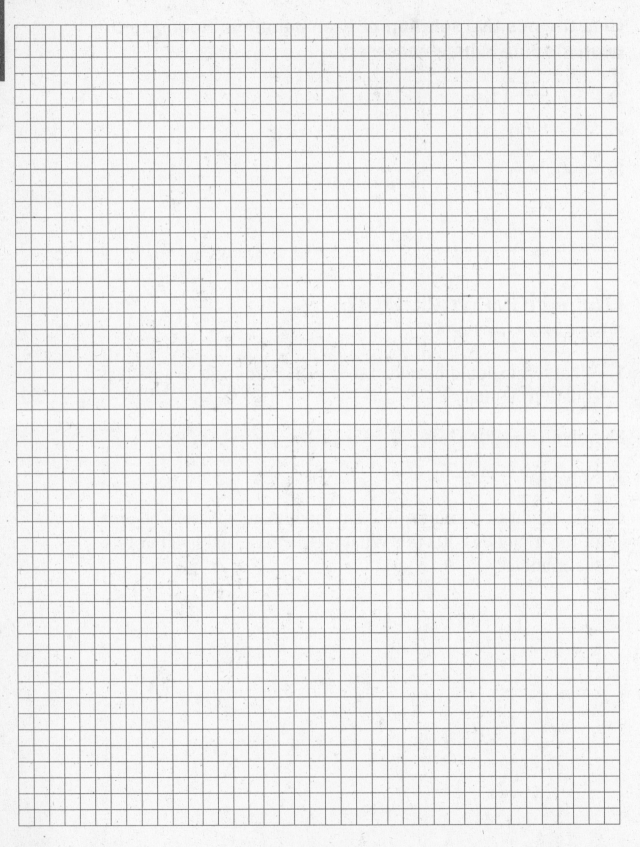

Algebra 1
Chapter 12 Resource Book

NAME _____ DATE _____

Visual Approach Lesson Opener

For use with pages 738–744

Square the length of each side of the right triangle shown in the picture. Add the squares of the lengths of the two shorter sides. Tell how this sum compares to the square of the length of the longest side.

1.
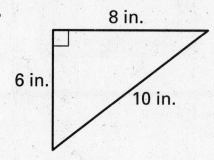

2.
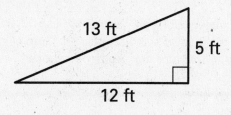

3.
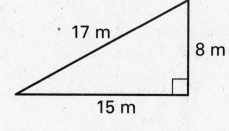

4.
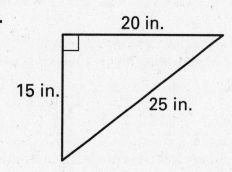

5.
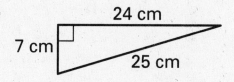

6.
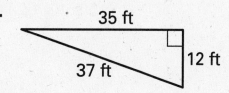

7. Make a conjecture about the relationship between the lengths of the sides of a right triangle.

NAME _____ DATE _____

Graphing Calculator Activity

For use with pages 738–744

GOAL **To write a program to test if a triangle is a right triangle**

In Lesson 12.5, you will learn about the Pythagorean theorem. The Pythagorean theorem can be used to determine if a triangle is a right triangle. Given a triangle with sides a, b, and c, according to the Pythagorean theorem $a^2 + b^2 = c^2$.

A graphing calculator can be programmed to use the Pythagorean theorem to test if a triangle is a right triangle. Graphing calculators use step-by-step instructions to perform operations.

Activity

The algorithm below shows the steps needed when writing a program that uses the Pythagorean theorem to determine if a triangle is a right triangle.

1 Enter the values for a, b, and c, where c is the largest side.

2 If $a^2 + b^2 = c^2$ is a true statement, the sides do determine a right triangle.

3 If $a^2 + b^2 = c^2$ is not a true statement, the sides do not determine a right triangle.

4 Follow your graphing calculator's procedure to enter a program. Run the program to decide whether each set of three sides determines a right triangle.

a. 7, 11, 15   **b.** 14, 48, 50   **c.** 6.8, 7.6, 13.4

Exercises

Use your graphing calculator program to decide whether the set of three sides determines a right triangle.

1. 6, 8, 10   **2.** 23, 27, 67   **3.** 15, 28, 42

4. 12, 34, 37   **5.** 0.8, 1.5, 1.7   **6.** 8.9, 11.2, 13.6

See page 69 for keystrokes.

Graphing Calculator Activity

For use with pages 738–744

TI-82

PROGRAM: RIGHTTRI
: Prompt A, B, C
: If $A^2 + B^2 = C^2$
: Then
: Disp "RIGHT TRIANGLE"
: Else
: Disp "NOT A RIGHT"
: Disp "TRIANGLE"
: End

TI-83

PROGRAM: RIGHTTRI
: Prompt A, B, C
: If $A^2 + B^2 = C^2$
: Then
: Disp "RIGHT TRIANGLE
: Else
: Disp "NOT A RIGHT"
: Disp "TRIANGLE"
: End

SHARP EL-9600c

RIGHTTRI
Input A
Input B
Input C
If $A^2 + B^2 = C^2$ Goto 1
If $A^2 + B^2 \neq C^2$ Goto 2
Label 1
Print "RIGHT TRIANGLE"
End
Label 2
Print "NOT A RIGHT TRIANGLE"
End

CASIO CFX-9850GA PLUS

RIGHTTRI
"A = "? → A↵
"B = "? → B↵
"C = "? → C↵
If $A^2 + B^2 = C^2$↵
Then "RIGHT TRIANGLE"↵
Else↵
"NOT A RIGHT TRIANGLE"
IfEnd↵

NAME _____ DATE _____

Practice A

For use with pages 738–744

Name the legs and the hypotenuse of the right triangle.

1.

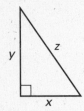

2.

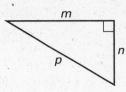

3.

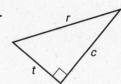

Find the missing length of the right triangle if *a* and *b* are the lengths of the legs and *c* is the length of the hypotenuse.

4.

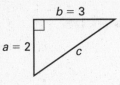

5.

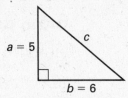

6.

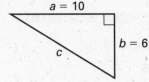

7. $a = 4, b = 6$

8. $a = 7, b = 3$

9. $a = 6, b = 6$

10. $a = 12, b = 9$

11. $a = 6, b = 8$

12. $a = 5, b = 12$

Find each missing length.

13.

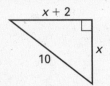

14.

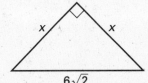

15.

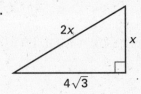

Determine whether the given lengths are sides of a right triangle. Explain your reasoning.

16. 2, 2, 4

17. 6, 9, 12

18. 10, 15, 20

19. 10, 24, 26

20. $5, 5, 5\sqrt{2}$

21. 30, 40, 50

22. *Baseball*   The infield of a baseball field is a square. The distance between consecutive bases is 90 feet. How far is it from first base to third base?

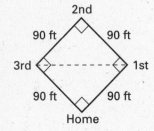

23. *Basketball*   A basketball court is a rectangle. If the court measures 60 feet by 40 feet, what is the length of the diagonal from one corner of the court to the opposite corner?

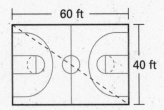

NAME _____ DATE _____

Practice B

For use with pages 738–744

Find the missing side length of the right triangle if *a* and *b* are the lengths of the legs and *c* is the length of the hypotenuse.

1.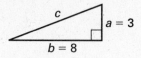

2.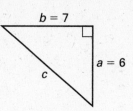

3.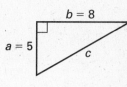

4. $a = 6, b = 7$

5. $a = 8, b = 12$

6. $a = 5, b = 10$

7. $a = 11, b = 13$

8. $a = 10, b = 24$

9. $a = 15, b = 18$

Find each missing length.

10.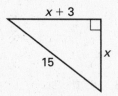

11.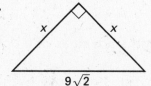

12.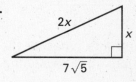

Determine whether the given lengths are sides of a right triangle. Explain your reasoning.

13. 4, 4, 8

14. 9, 12, 15

15. 12, 18, 24

16. 0.3, 0.4, 0.5

17. $8, 8, 8\sqrt{2}$

18. $\frac{3}{2}, \frac{4}{2}, \frac{5}{2}$

State the hypothesis and the conclusion of the statement.

19. If a quadrilateral is a square, then it is a rectangle.

20. If a triangle is equilateral, then all of the side lengths are congruent.

21. If today is February 29, then it is a leap year.

22. ***Indirect Measurement*** You are trying to determine the distance across a lake. You lay out posts at *A*, *B* and *C* so that angle *B* is a right angle (see figure). You measure and find the length *AB* is 22 feet and *CB* is 34 feet. How wide is the lake from *A* to *C*?

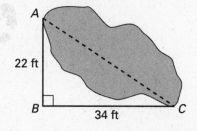

23. ***Volleyball*** You are setting up a volleyball net. To keep each pole standing straight, you use two ropes and two stakes as shown. How long is each piece of rope?

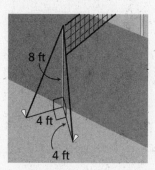

NAME _____ DATE _____

Practice C

For use with pages 738–744

Find the missing length of the right triangle if _a_ and _b_ are the lengths of the legs and _c_ is the length of the hypotenuse.

1. $a = 3, b = 8$

2. $a = 5, b = 9$

3. $a = 9, b = 9$

4. $a = 8, b = 12$

5. $a = 11, b = 12$

6. $a = 15, b = 20$

7. $a = \sqrt{2}, b = \sqrt{7}$

8. $a = \frac{3}{4}, b = 1$

9. $a = \frac{3}{5}, b = \frac{4}{5}$

Find each missing length.

10.
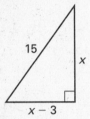

11.
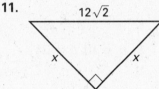

12.
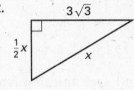

Determine whether the given lengths are sides of a right triangle. Explain your reasoning.

13. 6, 6, 10

14. 12, 16, 20

15. 15, 18, 21

16. 0.1, 0.2, 0.3

17. 12, 12, $12\sqrt{2}$

18. $\frac{6}{5}, \frac{8}{5}, 2$

State the hypothesis and the conclusion of the statement.

19. If you visit the school's website, then you will see a picture of the math team.

20. If all of the angles of a triangle measure less than 90°, then it is an acute triangle.

21. If the sides of the triangle measure 6 centimeters, 8 centimeters, and 12 centimeters, then the triangle is scalene.

22. *Shortest Route* You are traveling from Monroe to Harrisville. You can avoid the city traffic by staying on the roads shown. If you do travel straight through the city, how many miles will you save?

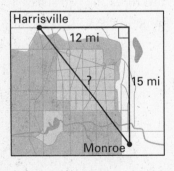

23. *Television Tower* The television tower is 110 feet tall. The four support braces are 24 feet out from the base of the tower. How long are the four braces?

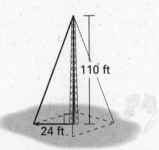

24. *Football* A football field is a rectangle that is 100 yards by 60 yards. What is the length of the diagonal from one corner of the field to the opposite corner?

NAME _____ DATE _____

Reteaching with Practice

For use with pages 738–744

GOAL Use the Pythagorean theorem and its converse and use the Pythagorean theorem in real-life problems

---

VOCABULARY

In a right trangle, the **hypotenuse** is the side opposite the right angle; the other two sides are the **legs.**

The **Pythagorean theorem** states that if a triangle is a right triangle, then the sum of the squares of the lengths of the legs a and b equals the square of the length of the hypotenuse c, or $a^2 + b^2 = c^2$.

In a statement of the form "If p, then q," p is the **hypothesis** and q is the **conclusion.** The **converse** of the statement "If p, then q" is the related statement "If q, then p."

The **converse of the Pythagorean theorem** states that if a triangle has side lengths a, b, and c such that $a^2 + b^2 = c^2$, then the triangle is a right triangle.

EXAMPLE 1 *Using the Pythagorean Theorem*

A right triangle has one leg that is 1 inch longer than the other leg. The hypotenuse is 5 inches. Find the lengths of the legs.

SOLUTION

Sketch a right triangle and label the sides. Let x be the length of the shorter leg. Use the Pythagorean theorem to solve for x.

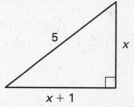

$$a^2 + b^2 = c^2 \qquad \text{Write Pythagorean theorem.}$$
$$x^2 + (x + 1)^2 = 5^2 \qquad \text{Substitute for } a, b, \text{ and } c.$$
$$x^2 + x^2 + 2x + 1 = 25 \qquad \text{Simplify.}$$
$$2x^2 + 2x - 24 = 0 \qquad \text{Write in standard form.}$$
$$2(x + 4)(x - 3) = 0 \qquad \text{Factor.}$$
$$x = -4 \text{ or } x = 3 \qquad \text{Zero-product property.}$$

Distance is positive. The sides have lengths 3 inches and $3 + 1 = 4$ inches.

LESSON 12.5 CONTINUED

Reteaching with Practice

For use with pages 738–744

Exercises for Example 1

Use the Pythagorean theorem to find the missing length of the right triangle.

1.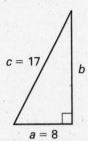

$c = 17$
b
$a = 8$

2.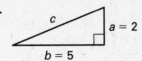

c
$a = 2$
$b = 5$

3.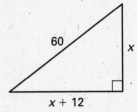

60
x
$x + 12$

EXAMPLE 2 *Determining Right Triangles*

Determine whether the given lengths are sides of a right triangle.

a. 2.5, 6, 6.5   **b.** 10, 24, 25

SOLUTION

Use the converse of the Pythagorean theorem.

a. The lengths are sides of a right triangle because

$2.5^2 + 6^2 = 6.25 + 36 = 42.25 = 6.5^2$.

b. The lengths are not sides of a right triangle because

$10^2 + 24^2 = 100 + 576 = 676 \neq 25^2$.

Exercises for Example 2

Determine whether the given lengths are sides of a right triangle.

4. 8, 15, 17   **5.** 3, 6, 7   **6.** 9, 40, 41

Algebra 1
Chapter 12 Resource Book

NAME _____ DATE _____

Quick Catch-Up for Absent Students
For use with pages 737–744

The items checked below were covered in class on (date missed) _____

Activity 12.5: Investigating the Pythagorean Theorem (p. 737)

_____ **Goal:** Determine if the Pythagorean Theorem is true only for right triangles.

Lesson 12.5: The Pythagorean Theorem and Its Converse

_____ **Goal 1:** Use the Pythagorean Theorem and its converse. (pp. 738–739)

Material Covered:

     _____ Example 1: Using the Pythagorean Theorem

     _____ Example 2: Using the Pythagorean Theorem

     _____ Student Help: Look Back

     _____ Student Help: Study Tip

     _____ Example 3: Determining Right Triangles

Vocabulary:

Pythagorean theorem, p. 738	hypotenuse, p. 738
legs of a right triangle, p. 738	hypothesis, p. 739
conclusion, p. 739	converse, p. 739

_____ **Goal 2:** Use the Pythagorean Theorem and its converse in real-life problems. (p. 740)

Material Covered:

     _____ Example 4: Using the Pythagorean Theorem

     _____ Example 5: Using the Pythagorean Converse

_____ Other (specify) _____

_____

Homework and Additional Learning Support

     _____ Textbook (specify) _pp. 741–744_____

_____

     _____ *Reteaching with Practice* worksheet (specify exercises)_____

     _____ *Personal Student Tutor* for Lesson 12.5

NAME _____ DATE _____

Real-Life Application: When Will I Ever Use This?

For use with pages 738–744

Kites

A kite is an object that is flown in the air at the end of a piece of string. The name comes from a graceful and soaring bird called a kite. Paper, cloth, plastic, and nylon are some of the materials from which kites are made. Kites are made in hundreds of sizes, shapes, and colors.

Kites are the oldest form of aircraft. They probably originated in China more than 2000 years ago. The Chinese military attached bamboo pipes to the kites. As the kites flew over the enemy, wind passed through the pipes, causing a whistling sound. The noise caused the enemy to panic and run. In 1752, Benjamin Franklin conducted the most famous kite experiment in history. A bolt of lightning struck a pointed wire fastened to the kite and traveled down the wet string to a key, causing a spark. In 1847, a kite helped pull a cable across the Niagara River between the United States and Canada. The cable was part of the river's first suspension bridge.

In Exercises 1-3, use the following information.

You and your friend each have a kite that you are going to fly. You have 150 feet of string and your friend has 100 feet of string.

1. You want to find how high your kite is in the air. First, you let the kite go as high as it can (until you run out of string). Next, you have your friend stand directly underneath your kite and you find that the distance between the two of you is 120 feet. Use the Pythagorean Theorem to find the height of your kite.

2. Your friend is flying her kite using only part of the string. You find that the distance between the two of you is 40 feet. You estimate that the height of the kite is 50 feet. Find the length of the string used by your friend.

3. Using your answer in Exercise 2, find the length of the string not used by your friend.

4. You see a woman flying a stunt kite with plastic hooks 140 feet up the string. She maneuvers the kite to pick up a stuffed toy monkey with the plastic hooks. She then drops the monkey. You estimate that the monkey fell 80 feet. Find the distance between the woman and where the monkey landed.

NAME _____ DATE _____

Math and History Application

For use with page 744

HISTORY   Pythagoras was born around 580 B.C. in Samos, Greece. He studied mathematics, music theory, and astronomy. Although the Pythagorean theorem is credited to Pythagoras, other cultures used the theorem many years before Pythagoras' time. The Chinese used the Pythagorean theorem to survey land, and the Egyptians used it to build pyramids, in each case, centuries before Pythagoras was born. Babylonian and Indian mathematicians used sets of three numbers where the sum of the squares of two is equal to the third. Lists of these triplets were found in the *Sulva Sultras,* which was written between 800 B.C. and 500 B.C. Trade was widespread between these areas so it is unknown where the theorem was first developed.

Pythagoras lived in Babylon for 20 years. During this time he studied and taught astronomy, mathematics, and astrology. He also traveled in Egypt and became acquainted with the mathematics of the area. These experiences probably led Pythagoras to record the theorem that bears his name.

MATH   Pythagoras returned to Italy and set up a secret society devoted to exploring the mysteries of numbers. He studied the relationship between numbers and musical harmonies. He developed theories for perfect, triangular, and square numbers. The Pythagorean theorem is very useful and has many applications.

1. You are building a tool shed. The framing for the floor measures 12 feet by 7 feet. What must the diagonal measure be to ensure the floor is square? (Carpenters consider the framing square when the angles measure 90°).

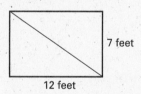

7 feet

12 feet

2. You and a friend are participating in an orienteering competition. Your friend travels off course and is now 600 meters from the checkpoint. You are 400 meters from the checkpoint. How far are you from your friend?

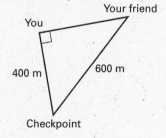

Your friend

You

400 m   600 m

Checkpoint

12.5

NAME _____ DATE _____

Challenge: Skills and Applications

For use with pages 738–744

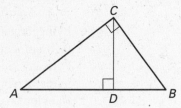

In Exercises 1 and 2, find the value of *x* and the lengths of sides AB, BC, and AC of triangle ABC.

1. $AD = 16$; $CD = 4x$; $DB = 3x$; $AC = 20$

2. $AD = 12x$; $CD = 5x$; $DB = 25$; $BC = 65$

In Exercises 3 and 4, use the following information.

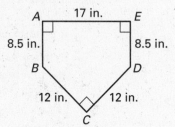

The Official Baseball Rule Book calls for home plate to have the dimensions shown in the diagram at the right, with right angles where they are indicated.

3. Use the Pythagorean Theorem to find *BD*. Give the length to the nearest tenth of an inch and to the nearest hundredth of an inch.

4. What inconsistency do you notice in the Rule Book diagram?

In Exercises 5–7, use the rectangular box with edges of length a, b, and c.

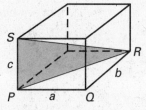

5. Let d = the length *PR*. Express d^2 in terms of *a*, *b*, and/or *c*.

6. Let e = the length of the diagonal *RS*. Express e^2 in terms of *d* and *c*.

7. Use the relationships from Exercises 5 and 6 to express e^2 in terms of *a*, *b*, and *c*.

In Exercises 8–10, use the diagram at the right.

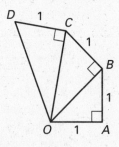

8. Find the length *OD*.

9. Suppose the diagram were continued to include 7 more right triangles, each with the shorter leg of length 1 and with the longer leg coinciding with the hypotenuse of the triangle before it. What would be the length of the longest hypotenuse in the new diagram?

10. Find a formula for the length of the hypotenuse of the *n*th triangle.

LESSON 12.5

Quiz 2

For use after Lessons 12.4–12.5

1. Find the term that should be added to $x^2 + 6x$ to create a perfect square trinomial. *(Lesson 12.4)*

2. Solve the equation by completing the square. *(Lesson 12.4)*

   $$x^2 + 12x = 13$$

Answers

1. _____

2. _____

3. _____

4. _____

5. _____

3. Find the missing length. *(Lesson 12.5)*

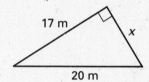

17 m

x

20 m

4. Determine whether 7, 9, and 11 are sides of a right triangle. Explain your reasoning. *(Lesson 12.5)*

5. A volleyball court is a rectangle 9 meters by 18 meters. What is the length of the diagonal of the court? *(Lesson 12.5)*

TEACHER'S NAME _____ CLASS _____ ROOM _____ DATE _____

Lesson Plan

1-day lesson (See *Pacing the Chapter,* TE pages 706C–706D)   **For use with pages 745–750**

GOALS **1. Find the distance between two points in a coordinate plane.**
2. Find the midpoint between two points in a coordinate plane.

State/Local Objectives _____

_____

✓ **Check the items you wish to use for this lesson.**

STARTING OPTIONS
____ Homework Check: TE page 741; Answer Transparencies
____ Warm-Up or Daily Homework Quiz: TE pages 745 and 743, CRB page 82, or Transparencies

TEACHING OPTIONS
____ Motivating the Lesson: TE page 746
____ Lesson Opener (Application): CRB page 83 or Transparencies
____ Examples 1–5: SE pages 745–747
____ Extra Examples: TE pages 746–747 or Transparencies
____ Closure Question: TE page 747
____ Guided Practice Exercises: SE page 748

APPLY/HOMEWORK
Homework Assignment
____ Basic   14–48 even, 49–53, 59–61, 66–72 even
____ Average   14–48 even, 49–53, 59–61, 66–72 even
____ Advanced   14–48 even, 49–53, 56–64, 66–72 even

Reteaching the Lesson
____ Practice Masters: CRB pages 84–86 (Level A, Level B, Level C)
____ Reteaching with Practice: CRB pages 87–88 or Practice Workbook with Examples
____ Personal Student Tutor

Extending the Lesson
____ Applications (Interdisciplinary): CRB page 90
____ Challenge: SE page 750; CRB page 91 or Internet

ASSESSMENT OPTIONS
____ Checkpoint Exercises: TE page 746–747 or Transparencies
____ Daily Homework Quiz (12.6): TE page 750, CRB page 94, or Transparencies
____ Standardized Test Practice: SE page 750; TE page 750; STP Workbook; Transparencies

Notes _____

_____

_____

Lesson Plan for Block Scheduling

Half-day lesson (See *Pacing the Chapter,* TE pages 706C–706D)   For use with pages 745–750

 GOALS 1. Find the distance between two points in a coordinate plane.
2. Find the midpoint between two points in a coordinate plane.

State/Local Objectives _____

_____

_____

✓ **Check the items you wish to use for this lesson.**

STARTING OPTIONS

____ Homework Check: TE page 741; Answer Transparencies
____ Warm-Up or Daily Homework Quiz: TE pages 745 and
       743, CRB page 82, or Transparencies

TEACHING OPTIONS

____ Motivating the Lesson: TE page 746
____ Lesson Opener (Application): CRB page 83 or Transparencies
____ Examples 1–5: SE pages 745–747
____ Extra Examples: TE pages 746–747 or Transparencies
____ Closure Question: TE page 747
____ Guided Practice Exercises: SE page 748

APPLY/HOMEWORK

Homework Assignment (See also the assignment for Lesson 12.5.)

____ Block Schedule: 14–48 even, 49–53, 56–61, 66–72 even

Reteaching the Lesson

____ Practice Masters: CRB pages 84–86 (Level A, Level B, Level C)
____ Reteaching with Practice: CRB pages 87–88 or Practice Workbook with Examples
____ Personal Student Tutor

Extending the Lesson

____ Applications (Interdisciplinary): CRB page 90
____ Challenge: SE page 750; CRB page 91 or Internet

ASSESSMENT OPTIONS

____ Checkpoint Exercises: TE pages 746–747 or Transparencies
____ Daily Homework Quiz (12.6): TE page 750, CRB page 94, or Transparencies
____ Standardized Test Practice: SE page 750; TE page 750; STP Workbook; Transparencies

Notes _____

_____

_____

CHAPTER PACING GUIDE	
Day	**Lesson**
1	12.1 (all); 12.2 (begin)
2	12.2 (end); 12.3 (begin)
3	12.3 (end); 12.4 (begin)
4	12.4 (end); 12.5 (begin)
5	12.5 (end); **12.6 (all)**
6	12.7 (all); 12.8 (all)
7	Review/Assess Ch. 12

Lesson 12.6

NAME _____ DATE _____

WARM-UP EXERCISES

For use before Lesson 12.6, pages 745–750

1. To get home from school Jan walks 1.5 miles south, then 0.75 miles west. To get home from school Celia walks 0.5 miles east from school, then 1.85 miles north. Whose home is farther from school?

··

DAILY HOMEWORK QUIZ

For use after Lesson 12.5, pages 737–744

1. Find the missing lengths of the right triangle if a and b are the lengths of the legs and c is the length of the hypotenuse.

   a. $a = 12, b = x, c = 25$

   b. $a = x + 2, b = x, c = \sqrt{13}$

2. Are 8, 12, and 17 lengths of a right triangle? Explain your reasoning.

3. State the hypothesis and the conclusion of the following statement. "If a right triangle has legs that are 3 inches and 4 inches long, then the hypotenuse is 5 inches long."

NAME _____ DATE _____

Application Lesson Opener

For use with pages 745–750

Use the figure at the right.

The distance between city *A* and city *C* is 12 miles and the distance between city *B* and city *C* is 9 miles.

1. Suggest a way to find the distance between city *A* and city *B*.

2. Use your suggestion to find the distance.

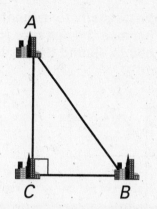

Use the figure at the right.

The distance between park *Z* and park *Y* is 5 miles and the distance between park *Y* and park *X* is 12 miles.

3. Suggest a way to find the distance between park *X* and park *Z*.

4. Use your suggestion to find the distance.

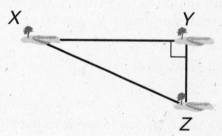

Use the figure at the right.

The distance between school *J* and school *K* is 8 miles and the distance between school *J* and school *L* is 15 miles.

5. Suggest a way to find the distance between school *K* and school *L*.

6. Use your suggestion to find the distance.

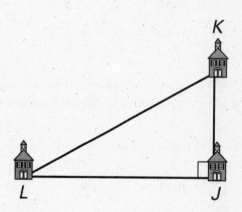

NAME _____ DATE _____

Practice A

For use with pages 745–750

Use the coordinate plane to estimate the distance between the two points. Then use the distance formula to find the distance between the points. Round the result to the nearest hundredth if necessary.

1.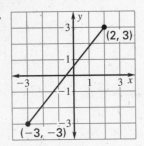

2.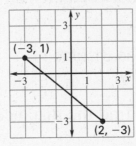

3.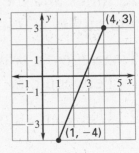

Find the distance between the two points. Round the result to the nearest hundredth if necessary.

4. $(1, 1), (4, 4)$

5. $(2, 5), (5, 1)$

6. $(0, 3), (2, 6)$

7. $(1, 6), (5, 1)$

8. $(-2, 8), (4, 0)$

9. $(3, -5), (-2, 0)$

10. $(-3, -5), (6, 5)$

11. $(8, 6), (-4, -3)$

12. $(-5, 2), (-2, 5)$

Use the distance formula to decide whether the three points are vertices of a right triangle.

13. $(1, 1), (4, 4), (4, 1)$

14. $(0, 6), (4, 6), (4, 2)$

15. $(-2, 6), (5, 3), (1, -2)$

16. $(3, -4), (-2, -1), (4, 6)$

Find the midpoint between the two points.

17. $(1, 1), (5, 5)$

18. $(2, 3), (4, 5)$

19. $(3, 0), (5, -4)$

20. $(-5, -2), (3, 6)$

21. $(0, -5), (3, -2)$

22. $(4, -1), (-1, 4)$

23. $(-3, -5), (-3, 2)$

24. $(2, -6), (-2, 3)$

25. $(-2, -4), (4, 6)$

Boston Suburbs **Use the map shown. Each side of each square is 4 kilometers. The points represent city locations.**

26. Use the distance formula to estimate the distance between Peabody and Bedford.

27. Use the distance formula to estimate the distance between Bedford and Hingham.

28. Use the distance formula to estimate the distance between Sudbury and Hingham.

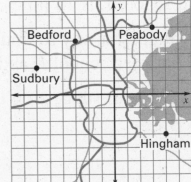

NAME _____ DATE _____

Practice B

For use with pages 745–750

Find the distance between the two points. Round the result to the nearest hundredth if necessary.

1. $(-2, -1), (2, 2)$

2. $(3, -6), (8, 6)$

3. $(3, 4), (2, 6)$

4. $(5, -4), (-2, -3)$

5. $(-2, -6), (-3, 4)$

6. $(4, -1), (5, 8)$

7. $(8, -2), (-3, 4)$

8. $(-1, -4), (-2, -5)$

9. $(-2, 3), (1, 5)$

10. $(3.2, -3), (4.8, -4)$

11. $\left(\frac{1}{4}, -\frac{1}{2}\right), (2, 2)$

12. $\left(\frac{2}{3}, \frac{1}{2}\right), \left(-\frac{5}{6}, \frac{1}{2}\right)$

Use the distance formula to decide whether the three points are vertices of a right triangle.

13. $(3, 5), (-2, 4), (2, 10)$

14. $(-4, 2), (-3, 2), (1, 6)$

15. $(-4, 1), (3, -2), (6, 5)$

16. $(4, -2), (3, 5), (-1, 0)$

17. $(1, 2), (-3, 6), (1, -5)$

18. $(-3, -2), (3, 4), (-8, 3)$

Find the midpoint between the two points.

19. $(4, 6), (2, 8)$

20. $(7, 1), (3, -5)$

21. $(-3, -6), (-1, -4)$

22. $(8, -5), (4, -3)$

23. $(0, 8), (-6, 4)$

24. $(3, 1), (-5, 4)$

25. $(-3, 1), (-5, 2)$

26. $(6, 3), (-1, -5)$

27. $(-2, 7), (-1, -8)$

28. $(-1, 2), (6, 1)$

29. $(-4, 3), (-1, 6)$

30. $(-5, -4), (2, -7)$

31. *Hockey*   A hockey player made a goal from the blue line. When he took the shot, he was on the blue line 10 feet from the center of the rink and the puck crossed the goal line at the center. How far did the puck travel?

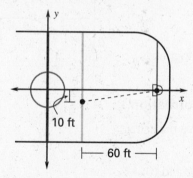

32. *Distance on a Map*   Each square on the grid superimposed on the map represents 80 miles by 80 miles. Use the map to estimate the distance from Honolulu to Hilo.

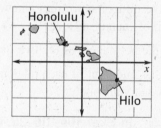

33. *Sales*   Use the midpoint formula to estimate the sales of a company in 1995, given the sales in 1990 and 2000. Assume the sales followed a linear pattern.

Year	1990	2000
Sales	$620,000	$830,000

34. *Rhombus*   A rhombus is a figure with two pairs of parallel sides all of equal length. Sketch the rhombus whose vertices are $(1, 0), (6, 0), (5, 3),$ and $(10, 3)$. Find the midpoint of each diagonal. What do you notice about these midpoints?

NAME _____ DATE _____

Practice C
For use with pages 745–750

Find the distance between the two points. Round the result to the nearest hundredth if necessary.

1. $(1, 5), (-3, 1)$ **2.** $(-2, 2), (2, 1)$ **3.** $(-3, -2), (4, 1)$

4. $(5, -2), (-1, 1)$ **5.** $(1, -7), (-2, 2)$ **6.** $(-4, 6), (1, -4)$

7. $(6, 9), (-3, 1)$ **8.** $(-7, -10), (-3, -6)$ **9.** $(-2.8, 7), (1.6, 2)$

10. $(-4.7, -5), (1.8, -2.6)$ **11.** $\left(\frac{1}{2}, \frac{1}{4}\right), \left(-\frac{1}{2}, \frac{5}{4}\right)$ **12.** $\left(\frac{2}{5}, \frac{1}{2}\right), \left(-\frac{3}{5}, -\frac{1}{2}\right)$

Use the distance formula to decide whether the three points are vertices of a right triangle.

13. $(4, 0), (2, 1), (-1, -5)$ **14.** $(4, 5), (1, 0), (-1, 2)$ **15.** $(1, -3), (3, 2), (-2, 4)$

16. $(-1, -1), (10, 7), (2, 18)$ **17.** $(2, 1), (4, 0), (5, 7)$ **18.** $(-7, -9), (3, -6), (-1, 7)$

Find the midpoint between the two points.

19. $(-4, 4), (2, 0)$ **20.** $(7, 0), (0, -10)$ **21.** $(-4, -8), (14, 6)$

22. $(10, -3), (6, -5)$ **23.** $(0, -5), (-6, -8)$ **24.** $(3.4, 6), (-2, 4)$

25. $(-2.8, 3), (-1, 3)$ **26.** $(4, 5.6), (-4, -1.8)$ **27.** $(-7, 2), (-5, -8)$

28. $\left(3\frac{1}{2}, -2\right), \left(-\frac{3}{4}, 6\right)$ **29.** $\left(-2\frac{3}{4}, \frac{1}{2}\right), \left(1\frac{1}{4}, 3\frac{1}{2}\right)$ **30.** $\left(\frac{4}{5}, -5\right), (4, -7)$

Geometry   **In Exercises 31–35, use the diagram at the right.**

31. Find the length of each side of the parallelogram.

32. Find the midpoint of each side of the parallelogram.

33. Join the midpoints to form a new quadrilateral. Find the length of each of its sides.

34. Find the perimeters of the two quadrilaterals.

35. Find the midpoint of each diagonal of the original parallelogram. What can you conclude?

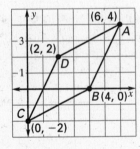

Trapezoids   **In Exercises 36–38, use the following information.**

A trapezoid is isosceles if its two opposite nonparallel sides have the same length.

36. Sketch the polygon whose vertices are $(1, 1), (5, 9), (2, 8),$ and $(0, 4)$.

37. Show that it is a trapezoid by showing that two of the sides are parallel.

38. Use the distance formula to show that the trapezoid is isosceles.

NAME _____ DATE _____

Reteaching with Practice

For use with pages 745–750

GOAL Find the distance between two points in a coordinate plane and find the midpoint between two points in a coordinate plane

> ### VOCABULARY
>
> The **midpoint** of a line segment is the point on the line segment that is equidistant from its endpoints.

EXAMPLE 1 *Finding the Distance Between Two Points*

Find the distance between $(-1, 2)$ and $(3, 7)$ using the distance formula.

$$d = \sqrt{(x_2 - x_1)^2 + (y_2 - y_1)^2}$$ Write distance formula.

$$= \sqrt{(-1 - 3)^2 + (2 - 7)^2}$$ Substitute.

$$= \sqrt{41}$$ Simplify.

$$\approx 6.40$$ Use a calculator.

Exercises for Example 1

Find the distance between the two points. Round the result to the nearest hundredth if necessary.

1. $(0, 4), (-3, 0)$ **2.** $(2, 3), (4, 5)$ **3.** $(-4, 2), (1, 4)$

EXAMPLE 2 *Applying the Distance Formula*

From your home, you ride your bicycle 5 miles north, then 12 miles east. How far are you from your home?

SOLUTION

You can superimpose a coordinate plane on a diagram of your bicycle trip. You start at the point $(0, 0)$ and finish at the point $(12, 5)$. Use the distance formula.

$$d = \sqrt{(12 - 0)^2 + (5 - 0)^2}$$

$$= \sqrt{144 + 25}$$

$$= \sqrt{169}$$

$$= 13$$

You are 13 miles from your home.

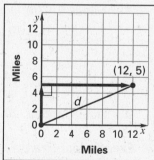

NAME _____ DATE _____

Reteaching with Practice

For use with pages 745–750

Exercise for Example 2

4. Rework Example 2 if you ride 8 miles east and 15 miles south.

EXAMPLE 3 *Finding the Midpoint Between Two Points*

Find the midpoint between $(-8, -4)$ and $(2, 0)$.

SOLUTION

Use the midpoint formula for the points (x_1, y_1) and (x_2, y_2): $\left(\dfrac{x_1 + x_2}{2}, \dfrac{y_1 + y_2}{2} \right)$.

$$\left(\frac{-8 + 2}{2}, \frac{-4 + 0}{2} \right) = \left(\frac{-6}{2}, \frac{-4}{2} \right) = (-3, -2)$$

The midpoint is $(-3, -2)$.

Exercises for Example 3

Find the midpoint between the two points.

5. $(1, 3), (4, 5)$ **6.** $(6, 1), (-4, -1)$ **7.** $(6, 0), (0, 2)$

EXAMPLE 4 *Applying the Midpoint Formula*

You and a friend agree to meet halfway between your two towns, as shown on the coordinate system at the right. Find the location where you should meet.

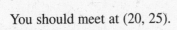

SOLUTION

The coordinates of your town are $(5, 10)$ and the coordinates of your friend's town are $(35, 40)$. Use the midpoint formula to find the point that is halfway between the two towns.

$$\left(\frac{5 + 35}{2}, \frac{10 + 40}{2} \right) = \left(\frac{40}{2}, \frac{50}{2} \right)$$
$$= (20, 25)$$

You should meet at $(20, 25)$.

Exercise for Example 4

8. Rework Example 4 if the coordinates of your town are $(0, 35)$ and the coordinates of your friend's town are $(30, 15)$.

NAME _____ DATE _____

Quick Catch-Up for Absent Students

For use with pages 745–750

The items checked below were covered in class on (date missed) _____

Lesson 12.6: The Distance and Midpoint Formulas

_____ **Goal 1:** Find the distance between two points in a coordinate plane. (pp. 745–746)

Material Covered:

_____ Activity: Investigating Distance

_____ Example 1: Finding the Distance Between Two Points

_____ Example 2: Checking a Right Triangle

_____ Example 3: Applying the Distance Formula

Vocabulary:

distance formula, p. 745

_____ **Goal 2:** Find the midpoint between two points in a coordinate plane. (p. 747)

Material Covered:

_____ Example 4: Finding the Midpoint Between Two Points

_____ Example 5: Applying the Midpoint Formula

Vocabulary:

midpoint between two points, p. 747   midpoint formula, p. 747

_____ Other (specify) _____

_____

Homework and Additional Learning Support

_____ Textbook (specify) _pp. 748–750_____

_____

_____ *Reteaching with Practice* worksheet (specify exercises)_____

_____ *Personal Student Tutor* for Lesson 12.6

NAME _____ DATE _____

Interdisciplinary Application

For use with pages 745–750

Model Planes

SCIENCE Your science class is building model planes. The planes can be made of any combination of wood, paper, glue, paper clips, or styrofoam™. Any gliding design is acceptable, but no propelling devices are allowed. Your grade will be determined based on the design of the plane and the distance traveled by the plane.

Every plane will be thrown from a window by your teacher. The window is 33 feet high. You build two designs at home. The first is made of styrofoam™, while the second is made from balsa wood. You test out your planes by throwing them from a window which is 22 feet high and measure the distance they travel. The styrofoam™ design travels 26 feet from the house. The wood design travels 32 feet from the house.

1. Refer to the diagram at the right. Determine the distance traveled from the window to the ground by the styrofoam™ plane.

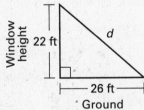

Window height 22 ft — *d* — 26 ft — Ground

2. Sketch a diagram similar to the one in Exercise 1 that represents the balsa wood plane.

3. Determine the distance traveled by the balsa wood plane.

In Exercises 4 and 5, use the following information

After testing your planes, you decide to make some modifications to the wing design of the balsa wood plane. You redesign your plane with curved wings to increase the lift of the plane.

4. Your teacher throws the redesigned plane from the window 33 feet above ground and it lands 64 feet from the building. What is the air distance traveled?

5. If all other conditions were the same at school and your house, did the new wing design cause the plane to travel further, or was it due to the greater height of the building?

NAME _____ DATE _____

Challenge: Skills and Applications

For use with pages 745–750

In Exercises 1–5, find the value of *k* to fit the given conditions.

1. $\left(\dfrac{5}{2}, \dfrac{1}{2}\right)$ is the midpoint of $(8, 5)$ and $(k, -4)$.

2. $(3k, -5)$ is the midpoint of $(k, -8)$ and $(3, -2)$.

3. The distance between $(k, -3)$ and $(4, -1)$ is $\sqrt{85}$.

4. The distance between $(7, k)$ and $(-3, 5)$ is $\sqrt{149}$.

5. The distance between $(2, k)$ and $(k, -5)$ is $\sqrt{137}$.

In Exercises 6–8, find the values of *a* and *b* to fit the given conditions.

6. $\left(5, -\dfrac{5}{2}\right)$ is the midpoint of (a, b) and $(7, 4)$.

7. $\left(0, \dfrac{b-3}{2}\right)$ is the midpoint of $(3a, -5)$ and $(2b, -a)$.

8. $\left(-19, \dfrac{b}{2}\right)$ is the midpoint of $\left(-12a, \dfrac{2}{3}\right)$ and $\left(5b, \dfrac{a}{3}\right)$.

In Exercises 9–12, use the following information.

At 12:00 noon, Isabel started walking east from her house along a straight road at 2 miles per hour. Her sister left the house to go running at 1:00 P.M., heading south at 8 miles per hour, also along a straight road. Let the house be at the origin of a coordinate system with north up.

9. Give the coordinates of Isabel's position *t* hours after noon. Give the coordinates of her sister's position at the same time.

10. Use the distance formula to write an expression for the distance between Isabel and her sister *t* hours after noon, assuming $t > 1$. Simplify the expression.

11. Find the distance between the sisters at 2:00 P.M.

12. At what time are the sisters 5 miles apart?

TEACHER'S NAME _____ CLASS _____ ROOM _____ DATE _____

Lesson Plan

1-day lesson (See *Pacing the Chapter,* TE pages 706C–706D)   **For use with pages 751–757**

GOALS 1. **Use the sine, cosine, and tangent of an angle.**
            2. **Use trigonometric ratios in real-life problems.**

State/Local Objectives _____

_____

✓ **Check the items you wish to use for this lesson.**

STARTING OPTIONS
____ Homework Check: TE page 748; Answer Transparencies
____ Warm-Up or Daily Homework Quiz: TE pages 752 and 750, CRB page 94, or Transparencies

TEACHING OPTIONS
____ Concept Activity: SE page 751; CRB page 95 (Activity Support Master)
____ Lesson Opener (Calculator): CRB page 96 or Transparencies
____ Graphing Calculator Activity with Keystrokes: CRB pages 97–98
____ Examples 1–3: SE pages 752–754
____ Extra Examples: TE pages 753–754 or Transparencies
____ Closure Question: TE page 754
____ Guided Practice Exercises: SE page 755

APPLY/HOMEWORK
Homework Assignment
____ Basic   10–19, 22, 24, 28–44 even
____ Average   10–19, 22, 24, 28–44 even
____ Advanced 10–19, 22–27, 28–44 even

Reteaching the Lesson
____ Practice Masters: CRB pages 99–101 (Level A, Level B, Level C)
____ Reteaching with Practice: CRB pages 102–103 or Practice Workbook with Examples
____ Personal Student Tutor

Extending the Lesson
____ Cooperative Learning Activity: CRB page 105
____ Applications (Real-Life): CRB page 106
____ Challenge: SE page 757; CRB page 107 or Internet

ASSESSMENT OPTIONS
____ Checkpoint Exercises: TE pages 753–754 or Transparencies
____ Daily Homework Quiz (12.7): TE page 757, CRB page 110, or Transparencies
____ Standardized Test Practice: SE page 757; TE page 757; STP Workbook; Transparencies

Notes _____

_____

_____

Lesson Plan for Block Scheduling

Half-day lesson (See *Pacing the Chapter*, TE pages 706C–706D)   For use with pages 751–757

GOALS
1. Use the sine, cosine, and tangent of an angle.
2. Use trigonometric ratios in real-life problems.

State/Local Objectives _____

_____

_____

✓ **Check the items you wish to use for this lesson.**

STARTING OPTIONS

_____ Homework Check: TE page 748; Answer Transparencies

_____ Warm-Up or Daily Homework Quiz: TE pages 752 and
           750, CRB page 94, or Transparencies

TEACHING OPTIONS

_____ Concept Activity: SE page 751; CRB page 95 (Activity Support Master)

_____ Lesson Opener (Calculator): CRB page 96 or Transparencies

_____ Graphing Calculator Activity with Keystrokes: CRB pages 97–98

_____ Examples 1–3: SE pages 752–754

_____ Extra Examples: TE pages 753–754 or Transparencies

_____ Closure Question: TE page 754

_____ Guided Practice Exercises: SE page 755

APPLY/HOMEWORK

Homework Assignment **(See also the assignment for Lesson 12.8.)**

_____ Block Schedule: 10–19, 22, 24, 28–44 even

Reteaching the Lesson

_____ Practice Masters: CRB pages 99–101 (Level A, Level B, Level C)

_____ Reteaching with Practice: CRB pages 102–103 or Practice Workbook with Examples

_____ Personal Student Tutor

Extending the Lesson

_____ Cooperative Learning Activity: CRB page 105

_____ Applications (Real-Life): CRB page 106

_____ Challenge: SE page 757; CRB page 107 or Internet

ASSESSMENT OPTIONS

_____ Checkpoint Exercises: TE pages 753–754 or Transparencies

_____ Daily Homework Quiz (12.7): TE page 757, CRB page 110, or Transparencies

_____ Standardized Test Practice: SE page 757; TE page 757; STP Workbook; Transparencies

Notes _____

_____

_____

CHAPTER PACING GUIDE	
Day	**Lesson**
1	12.1 (all); 12.2 (begin)
2	12.2 (end); 12.3 (begin)
3	12.3 (end); 12.4 (begin)
4	12.4 (end); 12.5 (begin)
5	12.5 (end); 12.6 (all)
6	**12.7 (all)**; 12.8 (all)
7	Review/Assess Ch. 12

Lesson 12.7

NAME _____ DATE _____

WARM-UP EXERCISES

For use before Lesson 12.7, pages 751–757

1. Which equation is equivalent to $0.2458 = \dfrac{20}{s}$?

a. $s = \dfrac{0.2458}{20}$ **b.** $s = \dfrac{20}{0.2458}$ **c.** $0.2458 = \dfrac{s}{20}$

Solve to the nearest hundredth.

2. $0.3558 = \dfrac{12}{q}$ **3.** $0.7288 = \dfrac{b}{252}$

DAILY HOMEWORK QUIZ

For use after Lesson 12.6, pages 745–750

1. Find the distance to the nearest tenth between the points $(4, 8)$ and $(-3, 5)$.

2. Graph $(1, 3)$, $(-2, 3)$, and $(1, 1)$. Decide whether they are vertices of a right triangle.

3. Find the midpoint of $(-9, -1)$ and $(4, 11)$.

4. Each side of a square in the coordinate plane that is superimposed on the map represents 0.5 mile. Estimate the distance between school and home.

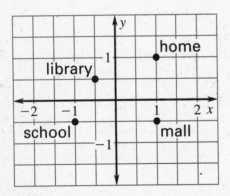

Algebra 1
Chapter 12 Resource Book

NAME _____ DATE _____

Activity Support Master

For use with page 751

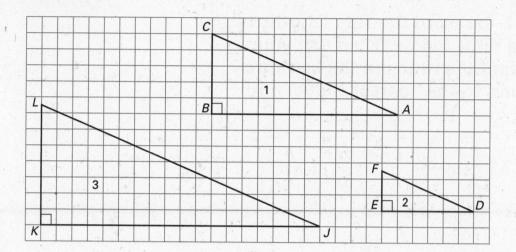

	Shorter leg	Longer leg	Hypotenuse	Shorter leg / Hypotenuse	Longer leg / Hypotenuse	Shorter leg / Longer leg
Δ1						
Δ2						
Δ3						

Cut strip out and use it to measure the hypotenuse.

NAME _____ DATE _____

Calculator Activity Lesson Opener

For use with pages 752–757

In Questions 1–4, record the side lengths of the
triangle in the table below. Then use your
calculator to find the ratios in the last three
columns. Round to the nearest thousandth.

	a	b	c	$\dfrac{a}{b}$	$\dfrac{a}{c}$	$\dfrac{b}{c}$
1.						
2.						
3.						
4.						

1.

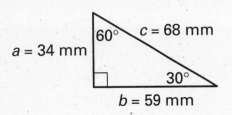

2.

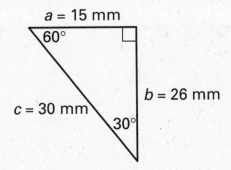

3.

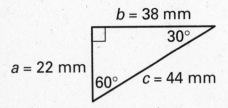

4.

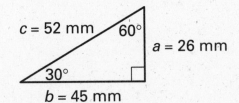

5. Look at the right triangles in Questions 1–4. What is
the same about the triangles? What is different?

6. What do you notice about the values in the last three
columns of the table?

NAME _____ DATE _____

Graphing Calculator Activity

For use with pages 752–757

GOAL **To discover how the sides of a right triangle relate to the trigonometric ratios sine, cosine, and tangent**

In Lesson 12.7, you will learn about the sine, cosine, and tangent. These are just names for the ratios of the sides of a right triangle. Your calculator can quickly calculate these ratios.

Activity

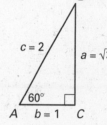

❶ Using the triangle at the right, write the ratio as a fraction and as a decimal.

    a. $\dfrac{a}{c}$ **b.** $\dfrac{b}{c}$ **c.** $\dfrac{a}{b}$

❷ Use your calculator to evaluate the following. Make sure your calculator is in degree mode.

    a. sin 60°   **b.** cos 60°   **c.** tan 60°

❸ How are the ratios in Step 1 related to your answers in Step 2?

❹ Use your calculator to evaluate the following. Make sure your calculator is in degree mode.

    a. cos 30°   **b.** sin 30°

❺ How are the sine and cosine of complementary angles (angles that add up to 90°) related? (*Hint:* Compare your answers to part (a) of Steps 2 and 4.)

Exercises

Using the triangle at the right, write the ratio as a fraction.

1. sin A   **2.** cos A   **3.** tan A

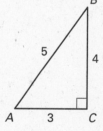

Write another trigonometric ratio that will give the same result as the one given.

4. sin 45°   **5.** cos 35°   **6.** sin 50°

7. Compare the tangents of two complementary angles. How are they related?

See page 98 for keystrokes.

Lesson 12.7

NAME _____ DATE _____

Graphing Calculator Activity

For use with pages 752–757

TI-82

MODE ▼ ▼ ▶ ENTER

2nd [QUIT]

SIN 60 ENTER

COS 60 ENTER

TAN 60 ENTER

COS 30 ENTER

SIN 30 ENTER

TI-83

MODE ▼ ▼ ▶ ENTER

2nd [QUIT]

SIN 60) ENTER

COS 60) ENTER

TAN 60) ENTER

COS 30) ENTER

SIN 30) ENTER

SHARP EL-9600c

2ndF [SET UP] [B] 1

2ndF [QUIT]

SIN 60 ENTER

COS 60 ENTER

TAN 60 ENTER

COS 30 ENTER

SIN 30 ENTER

CASIO CFX-9850GA PLUS

From the main menu, choose RUN.

SHIFT [SET UP] ▼ ▼ ▼ ▼ F1

EXIT

SIN 60 EXE

COS 60 EXE

TAN 60 EXE

COS 30 EXE

SIN 30 EXE

Lesson 12.7

Practice A
For use with pages 752–757

In Exercises 1–6, use △*MPQ* at the right.

1. Name the side opposite ∠*M*.

2. Name the side adjacent to ∠*M*.

3. Name the side opposite ∠*Q*.

4. Name the side adjacent to ∠*Q*.

5. Name the hypotenuse of △*MPQ*.

6. Name the acute angles of △*MPQ*.

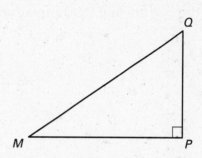

Find the sine, the cosine, and the tangent of ∠*A* and of ∠*B*.

7.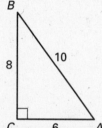

8.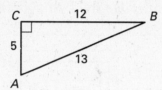

9.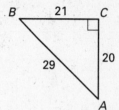

Find the missing lengths of the sides of the triangle. Round your answers to the nearest hundredth. Use the Pythagorean theorem to check.

10.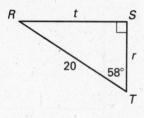

11.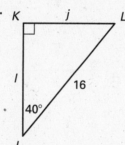

12.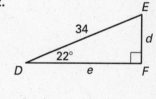

13. *Roof Construction*   A new home is being built. The roof is at a 35° angle as shown and the wooden runners that go from the corner of the house to the top of the roof are 24 feet. How high does the roof rise from the house?

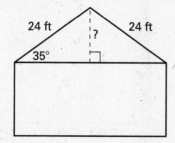

14. *Flying Kites*   You are flying a kite and have let out 100 feet of string. If the kite string makes a 45° angle with the horizontal, approximately how high is the kite in the sky? Assume that your height is 6 feet.

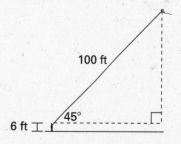

NAME _____ DATE _____

Practice B

For use with pages 752–757

Find the sine, the cosine, and the tangent of ∠A and of ∠B.

1.
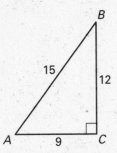

2.
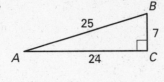

3.
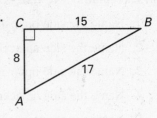

Find the missing lengths of the sides of the triangle. Round your answers to the nearest hundredth. Use the Pythagorean theorem to check.

4.
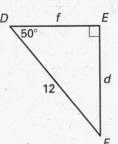

5.
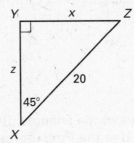

6.
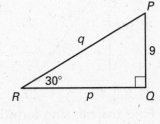

7.
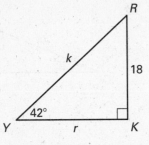

8.
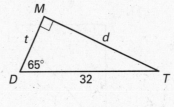

9.
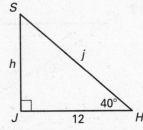

10. Ladder A ladder that leans against a house makes a 68° angle with the ground. If the ladder is 4 feet from the base of the house, how high up the house does the ladder reach?

11. Monument You are standing 50 feet from the base of a monument. The angle formed by the ground and your line of sight to the top of the monument is 42°. How tall is the monument? Assume that your height is 5 feet.

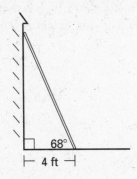

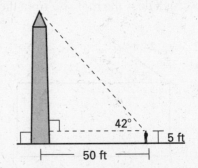

Find the sine, the cosine, and the tangent of ∠A and of ∠B.

1.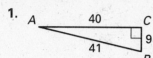

2.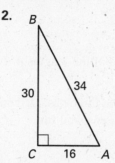

3.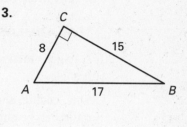

Find the missing lengths of the sides of the triangle. Round your answers to the nearest hundredth. Use the Pythagorean theorem to check.

4.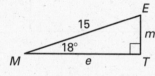

5.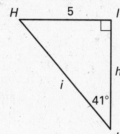

6.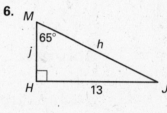

7.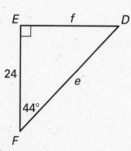

8.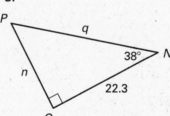

9.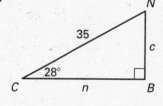

10. *Map Mileage*   Kanton is directly east of Weston. Weston is directly north of Audler. Find the distance from Weston to Audler. Find the distance from Audler to Kanton.

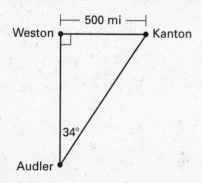

11. *Map Mileage*   You live in Atlanta, Georgia, and your friend lives in Memphis, Tennessee, 336 air miles away. You are each flying to Chicago, Illinois, to attend a science fair. How many air miles will each of you fly?

NAME _____ DATE _____

Reteaching with Practice

For use with pages 752–757

GOAL Use the sine, cosine, and tangent of an angle and use trigonometric ratios in real-life problems

---

VOCABULARY

A **trigonometric ratio** is a ratio of the lengths of two sides of a right triangle.

Sine, cosine, and **tangent** are the three basic trigonometric ratios. These ratios can be abbreviated as **sin, cos,** and **tan.**

EXAMPLE 1 *Finding Trigonometric Ratios*

For $\triangle ABC$, find the sine, the cosine, and the tangent of the angle.

a. $\angle A$

b. $\angle C$

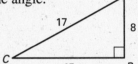

SOLUTION

a. For $\angle A$, the opposite side is 15 and the adjacent side is 8. The hypotenuse is 17.

$$\sin A = \frac{\text{opposite}}{\text{hypotenuse}} = \frac{15}{17}$$

$$\cos A = \frac{\text{adjacent}}{\text{hypotenuse}} = \frac{8}{17}$$

$$\tan A = \frac{\text{opposite}}{\text{adjacent}} = \frac{15}{8}$$

b. For $\angle C$, the opposite side is 8 and the adjacent side is 15. The hypotenuse is 17.

$$\sin C = \frac{\text{opposite}}{\text{hypotenuse}} = \frac{8}{17}$$

$$\cos C = \frac{\text{adjacent}}{\text{hypotenuse}} = \frac{15}{17}$$

$$\tan C = \frac{\text{opposite}}{\text{adjacent}} = \frac{8}{15}$$

Exercises for Example 1

Find the sine, the cosine, and the tangent of $\angle A$ and $\angle C$.

1.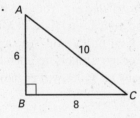

2.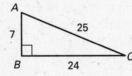

3.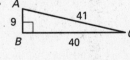

Algebra 1
Chapter 12 Resource Book

NAME _____ DATE _____

Reteaching with Practice

For use with pages 752–757

EXAMPLE 2 *Solving a Right Triangle*

For $\triangle XYZ$, $z = 6$ and the measure of $\angle X$ is 60°. Find the length of y.

SOLUTION

You are given the side adjacent to $\angle X$, and you need to find the length of the hypotenuse.

$\cos X = \dfrac{\text{adjacent}}{\text{hypotenuse}}$ Definition of cosine

$\cos 60° = \dfrac{6}{y}$ Substitute 6 for z and 60° for $\angle X$.

$0.5 = \dfrac{6}{y}$ Use a calculator or a table.

$\dfrac{6}{0.5} = y$ Solve for y.

$12 = y$ Simplify.

The length y is 12 units.

Exercises for Example 2

Find the missing lengths of the sides of the triangle. Round your answers to the nearest hundredth.

4.
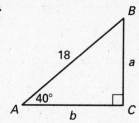

5.
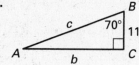

6.
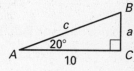

Lesson 12.7

Quick Catch-Up for Absent Students

For use with pages 751–757

The items checked below were covered in class on (date missed) _____

Activity 12.7: Investigating Similar Triangles (p. 751)

_____ **Goal:** Determine the relationship between the sides of similar right triangles.

Lesson 12.7: Trigonometric Ratios

_____ **Goal 1:** Use the sine, cosine, and tangent of an angle. (pp. 752–753)

Material Covered:

_____ Example 1: Finding Trigonometric Ratios

_____ Student Help: Trig Table

_____ Student Help: Study Tip

_____ Example 2: Solving a Right Triangle

Vocabulary:

trigonometric ratio, p. 752   sine, p. 752
cosine, p. 752   tangent, p. 752

_____ **Goal 2:** Use trigonometric ratios in real-life problems. (p. 754)

Material Covered:

_____ Example 3: Solving a Right Triangle

_____ Other (specify) _____

_____

Homework and Additional Learning Support

_____ Textbook (specify) _pp. 755–757_____

_____

_____ *Reteaching with Practice* worksheet (specify exercises)_____

_____ *Personal Student Tutor* for Lesson 12.7

NAME _____ DATE _____

Cooperative Learning Activity

For use with pages 752–757

GOAL **To indirectly measure the height of an object**

Materials: protractor, string, weight, straw, meter/yard stick, paper, pencil

The height of tall objects cannot be measured directly by using a meter stick or a tape measure. Indirect methods, including trigonometric ratios, must be used instead. In this activity, you and a partner will determine the height of a tall object using a homemade clinometer and your knowledge of trigonometric ratios.

Instructions

1 Make your homemade clinometer by attaching a piece of string to the center of a protractor (see diagram). Attach a small weight, such as a washer, to the end of the string. Tape a straw to the bottom of the protractor.

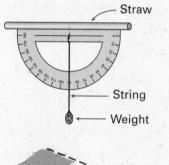

2 Locate an object that you would like to know the height of. It could be a tree or a tall building.

3 Choose a location some distance from the object. Measure the distance from your location to the object.

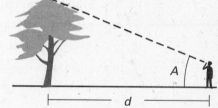

4 Sight the top of the object through the straw attached to the protractor. Your partner should then record the angle made by the string.

5 Use your knowledge of trigonometry to find the height of the object. (*Hint:* You should use the tangent ratio.)

$$\tan A = \frac{\text{opposite}}{\text{adjacent}}$$

Analyzing the Results

1. Can you find out the actual height of the object you measured? How close was your measurement to the actual height?

2. Where might this method of measuring the height of an object be useful in real life?

Lesson 12.7

NAME _____ DATE _____

Real-Life Application:
When Will I Ever Use This?

For use with pages 752–757

X-rays

X-rays are one of the most useful forms of energy. Wilhelm C. Roentgen, a German physicist, discovered them in 1895. Roentgen called the rays X-rays because he did not understand what they were at first. X is a scientific symbol for the unknown.

The wavelengths of X-rays are much shorter than the wavelengths of light. For this reason, X-rays can penetrate deeply into many substances that do not transmit light. The penetrating power and other characteristics of X-rays make them extremely useful in medicine, industry, and scientific research.

In medicine, X-rays are widely used to make radiographs (X-ray pictures) of the bones and internal organs of the body. Radiographs help physicians detect abnormalities and disease conditions, such as broken bones or lung disease, inside a patient's body. Dentists use X-rays to reveal cavities and impacted teeth.

In Exercises 1-3, use the following information.

An X-ray source is positioned over a piece of film, as illustrated at the right. The source is centered 36 inches away from the piece of film.

1. Find the approximate width of the film.

2. Use a trigonometric ratio to find how far the X-rays travel to reach the edges of the film.

3. Use the Pythagorean Theorem to check the result of Exercise 2.

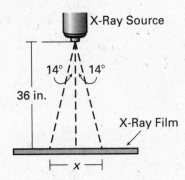

In Exercises 4 and 5, use the following information.

Another X-ray view requires the X-ray source to be positioned at an angle of 30° from the center line, yet still be a vertical distance of 36 inches from the film, as shown at the right.

4. How far from the center line should the X-ray source be positioned?

5. Use a trigonometric ratio to determine whether the X-ray source will be closer or farther than 36 inches from the center of the film. How much closer or farther than 36 inches will the X-ray source be?

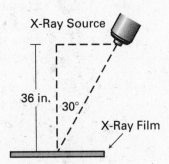

Challenge: Skills and Applications

For use with pages 752–757

In Exercises 1–4, you will explore an isosceles right triangle.

1. The sum of the measures of the angles of any triangle is 180°. The base angles of an isosceles triangle have the same measure. Use these facts to find the measure of a base angle of an isosceles right triangle.

2. Let a be the length of one leg of an isosceles right triangle. Write an expression in terms of a for the length of the hypotenuse.

3. Use the side lengths from Exercise 2 to find the following for a base angle of an isosceles right triangle. Express answers in simplest form.

   a. sine   **b.** cosine   **c.** tangent

4. Use a calculator to check that your results in Exercise 3 match the values for the sine, cosine, and tangent of the angle measure you found in Exercise 1.

5. When necessary, round answers to 4 decimal places.

   a. Find $\sin 30°$.   **b.** Find $\cos 30°$.

   c. Find $\tan 30°$.   **d.** Find $\dfrac{\sin 30°}{\cos 30°}$.

   e. What do you notice?

   f. Use the definitions of the trigonometric ratios to show your observation from part (e) is correct for any angle A.

In Exercises 6–8, use the following information. Round answers to the nearest hundredth of a foot.

As shown in the diagram, points D and E are along one edge of a canyon and points B and C are along the opposite, parallel edge of the canyon. The line containing points D and B is perpendicular to the edges of the canyon. Point A is 7.25 feet from point D. Point A is on line DB and on line EC and the measure of angle A is 25°.

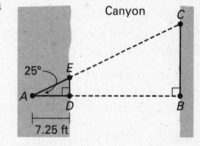

6. Find the distance from point A to point E.

7. Find the distance from point D to point E.

8. The distance between point B and point C is 13.5 feet. What is the distance across the canyon from point D to point B?

Lesson 12.7

TEACHER'S NAME _____ CLASS _____ ROOM _____ DATE _____

Lesson Plan

1-day lesson (See *Pacing the Chapter,* TE pages 706C–706D)   For use with pages 758–764

GOALS   1. **Use logical reasoning and proof to prove a statement is true.**
2. **Prove that a statement is false.**

State/Local Objectives _____

_____

✓ **Check the items you wish to use for this lesson.**

STARTING OPTIONS
____ Homework Check: TE page 755; Answer Transparencies
____ Warm-Up or Daily Homework Quiz: TE pages 758 and 757, CRB page 110, or Transparencies

TEACHING OPTIONS
____ Motivating the Lesson: TE page 759
____ Lesson Opener (Activity): CRB page 111 or Transparencies
____ Examples 1–5: SE pages 759–761
____ Extra Examples: TE pages 759–761 or Transparencies
____ Closure Question: TE page 761
____ Guided Practice Exercises: SE page 761

APPLY/HOMEWORK
Homework Assignment
____ Basic   14–21, 29, 35, 40, 45, 50; Quiz 3: 1–14
____ Average   14–21, 24, 25, 29, 35, 40, 45, 50; Quiz 3: 1–14
____ Advanced   14–30, 35, 40, 45, 50; Quiz 3: 1–14

Reteaching the Lesson
____ Practice Masters: CRB pages 112–114 (Level A, Level B, Level C)
____ Reteaching with Practice: CRB pages 115–116 or Practice Workbook with Examples
____ Personal Student Tutor

Extending the Lesson
____ Applications (Interdisciplinary): CRB page 118
____ Challenge: SE page 763; CRB page 119 or Internet

ASSESSMENT OPTIONS
____ Checkpoint Exercises: TE pages 759–761 or Transparencies
____ Daily Homework Quiz (12.8): TE page 764 or Transparencies
____ Standardized Test Practice: SE page 763; TE page 764; STP Workbook; Transparencies
____ Quiz (12.6–12.8): SE page 764

Notes _____

_____

_____

TEACHER'S NAME _____ CLASS _____ ROOM _____ DATE _____

Lesson Plan for Block Scheduling

Half-day lesson (See *Pacing the Chapter,* TE pages 706C–706D)   For use with pages 758–764

GOALS   1. Use logical reasoning and proof to prove a statement is true.
2. Prove that a statement is false.

State/Local Objectives _____

_____

_____

✓ **Check the items you wish to use for this lesson.**

STARTING OPTIONS

____ Homework Check: TE page 755; Answer Transparencies
____ Warm-Up or Daily Homework Quiz: TE pages 758 and
        757, CRB page 110, or Transparencies

TEACHING OPTIONS

____ Motivating the Lesson: TE page 759
____ Lesson Opener (Activity): CRB page 111 or Transparencies
____ Examples 1–5: SE pages 759–761
____ Extra Examples: TE pages 759–761 or Transparencies; Internet
____ Closure Question: TE page 761
____ Guided Practice Exercises: SE page 761

APPLY/HOMEWORK

Homework Assignment (See also the assignment for Lesson 12.7.)
____ Block Schedule: 14–21, 24, 25, 29, 35, 40, 45, 50; Quiz 3: 1–14

Reteaching the Lesson
____ Practice Masters: CRB pages 112–114 (Level A, Level B, Level C)
____ Reteaching with Practice: CRB pages 115–116 or Practice Workbook with Examples
____ Personal Student Tutor

Extending the Lesson
____ Applications (Interdisciplinary): CRB page 118
____ Challenge: SE page 763; CRB page 119 or Internet

ASSESSMENT OPTIONS

____ Checkpoint Exercises: TE pages 759–761 or Transparencies
____ Daily Homework Quiz (12.8): TE page 764 or Transparencies
____ Standardized Test Practice: SE page 763; TE page 764; STP Workbook; Transparencies
____ Quiz (12.6–12.8): SE page 764

Notes _____

_____

_____

CHAPTER PACING GUIDE	
Day	**Lesson**
1	12.1 (all); 12.2 (begin)
2	12.2 (end); 12.3 (begin)
3	12.3 (end); 12.4 (begin)
4	12.4 (end); 12.5 (begin)
5	12.5 (end); 12.6 (all)
6	12.7 (all); **12.8 (all)**
7	Review/Assess Ch. 12

Lesson 12.8

NAME _____ DATE _____

WARM-UP EXERCISES

For use before Lesson 12.8, pages 758–764

Which property of real numbers justifies each statement?

1. $3(a + b) = 3a + 3b$

2. If $x = 5$, then $x + 2 = 5 + 2$.

3. $4 + z = z + 4$

4. $(p + q) + 0 = p + q$

5. $5\left(\dfrac{1}{5}\right) = 1$

....

DAILY HOMEWORK QUIZ

For use after Lesson 12.7, pages 751–757

1. Find the sine, the cosine, and the tangent of $\angle A$ and $\angle B$.

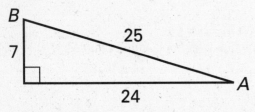

2. Find the missing lengths of the sides of the triangle. Round your answer to the nearest hundredth.

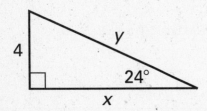

3. A string tied to a stake is attached to a kite. The string is 30 ft long. The angle between the ground and the string is 65°. How high is the kite?

Lesson 12.8

NAME ______________________ DATE ________

Available as a transparency

Activity Lesson Opener

For use with pages 758–764

SET UP: Work with a partner.

When you solve an equation, you can give an explanation for each step. An example is given below.

$x + 3 = 5$	Given: Write original equation.
$x + 3 - 3 = 5 - 3$	Subtract same number from each side.
$x + 0 = 2$	Subtract to simplify both sides.
$x = 2$	Add to simplify the left side.

The steps to solve an equation are shown on the left. Put the explanations shown on the right in the correct order.

Steps	Explanations
1. $2x - 3 = 5$	**A.** Add same number to both sides.
2. $2x - 3 + 3 = 5 + 3$	**B.** Divide to simplify both sides.
3. $2x + 0 = 8$	**C.** Add to simplify the left side.
4. $2x = 8$	**D.** Add to simplify both sides.
5. $\frac{2x}{2} = \frac{8}{2}$	**E.** Given
6. $x = 4$	**F.** Divide both sides by the same number.

State the basic axiom of algebra that is represented.

1. $n + 0 = n$

2. $7(m + n) = 7m + 7n$

3. $mn = nm$

4. $n(1) = n$

5. $3m + n = n + 3m$

6. $m + (-m) = 0$

7. $(5n)m = 5(nm)$

8. $n\left(\dfrac{1}{n}\right) = 1$

9. $(m + n) + p = m + (n + p)$

10. Copy and complete the proof of the statement: If $a = b$ and $c = d$, then $a + c = b + d$. Each variable represents any real number.

$a = b$ Given

$a + c = b + c$ a. _____

$c = d$ b. _____

$b + c = b + d$ c. _____

$a + c = b + d$ Substitution property of equality

11. Copy and complete the proof of the statement: $c\left(1 + \dfrac{1}{c}\right) = c + 1$. Let c represent any nonzero real number.

$c\left(1 + \dfrac{1}{c}\right) = c \cdot 1 + c \cdot \dfrac{1}{c}$ Distributive property

$c \cdot 1 = c$ a. _____

$c \cdot \dfrac{1}{c} = 1$ b. _____

$c\left(1 + \dfrac{1}{c}\right) = c + 1$ Substitution property of equality

Find a counterexample to show that the statement is *not* true.

12. If a and b are real numbers, then $|a + b| = |a| + |b|$.

13. If a and b are real numbers, then $\sqrt{a + b} = \sqrt{a} + \sqrt{b}$.

14. *Proof*   Prove that the diagonals of any rectangle are equal in length. You will need to use the distance formula and show $AC = BD$.

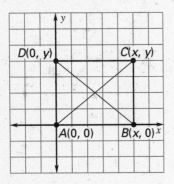

Lesson 12.8

NAME _____ DATE _____

Practice B

For use with pages 758–764

State the basic axiom of algebra that is represented.

1. $p + q = q + p$ **2.** $p + (-p) = 0$ **3.** $3(p + q) = 3p + 3q$

4. $(2p)q = 2(pq)$ **5.** $pq = qp$ **6.** $q + 0 = q$

7. $(pq)(1) = pq$ **8.** $pq\left(\dfrac{1}{pq}\right) = 1$ **9.** $(p + q) + r = p + (q + r)$

10. Copy and complete the proof of the statement: If $a = b$ and $-c = -d$,
then $a - c = b - d$. Each variable represents any real number.

$a = b$	Given
$a + (-c) = b + (-c)$	a. _____
$-c = -d$	b. _____
$b + (-c) = b + (-d)$	c. _____
$a + (-c) = b + (-d)$	Substitution property of equality
$a - c = b - d$	d. _____

Prove the theorem. Use the basic axioms of algebra and the definition of subtraction.

11. If p and q are real numbers, then $p - q = -q + p$.

12. If a, b, and c are real numbers, then $(a + b) + c = (a + c) + b$.

13. If m and n are real numbers, then $(m + n) - n = m$.

Find a counterexample to show that the statement is *not* true.

14. If a and b are real numbers, then $|a - b| = |a| - |b|$.

15. If a and b are real numbers, then $\sqrt{a - b} = \sqrt{a} - \sqrt{b}$.

16. *Proof* Prove that the diagonals of any rectangle are equal in length. You will need to use the distance formula and show $AC = BD$.

17. *Proof* Prove that the diagonals of any rectangle bisect one another. You will need to show that the midpoints of each diagonal are the same point.

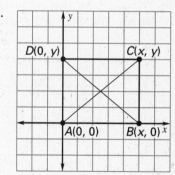

NAME _____ DATE _____

Practice C

For use with pages 758–764

State the basic axiom of algebra that is represented.

1. $ab + c = c + ab$

2. $b + (-b) = 0$

3. $4(a + c) = 4a + 4c$

4. $(cb)a = c(ba)$

5. $bc = cb$

6. $ac + 0 = ac$

7. $(ac)(1) = ac$

8. $bc\left(\dfrac{1}{bc}\right) = 1$

9. $(b + c) + a = b + (c + a)$

10. Copy and complete the proof of the statement: If $a = b$ and $-c = -d$,
then $a - c = b - d$. Each variable represents any real number.

$a = b$	**a.** _____
$a + (-c) = b + (-c)$	**b.** _____
$-c = -d$	**c.** _____
$b + (-c) = b + (-d)$	**d.** _____
$a + (-c) = b + (-d)$	**e.** _____
$a - c = b - d$	**f.** _____

**Prove the theorem. Use the basic axioms of algebra and the defini-
tion of subtraction.**

11. If a and b are real numbers, then $a - b = -b + a$.

12. If a, b, and c are real numbers, and $a + c = b + c$, then $a = b$.

13. If a and b are real numbers, then $(b + a) - b = a$.

Find a counterexample to show that the statement is *not* true.

14. If a, b, and c are real numbers and $a < b$, then $ac < bc$.

15. If a and b are real numbers and $a < b$, then $a^2 < b^2$.

16. *Proof* Prove that the diagonals of any isosceles trapezoid are equal in
length. You will need to use the distance formula and show $AC = BD$.

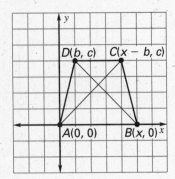

NAME _____ DATE _____

Reteaching with Practice

For use with pages 758–764

GOAL Use logical reasoning and proof to prove that a statement is true and prove that a statement is false

---

VOCABULARY

Postulates or **axioms** are rules in mathematics that we accept to be true without proof.

Theorems are other new statements that have to be proved.

A **conjecture** is a statement that is believed to be true but not yet proved.

In an **indirect proof,** or a **proof by contradiction,** you assume that the original statement is false. If this leads to an impossibility, then the original statement is true.

EXAMPLE 1 *Proving a Theorem*

Prove the cancellation property of addition: If $a + c = b + c$, then $a = b$.

SOLUTION

$a + c = b + c$	Given
$a + c + (-c) = b + c + (-c)$	Addition axiom of equality
$a + [c + (-c)] = b + [c + (-c)]$	Associative axiom of addition
$a + 0 = b + 0$	Inverse axiom of addition
$a = b$	Identity axiom of addition

Exercises for Example 1

Prove the theorem using the basic axioms of algebra.

1. If $ac = bc$ and $c \neq 0$, then $a = b$.   **2.** If $a - c = b - c$, then $a = b$.

EXAMPLE 2 *Finding a Counterexample*

Assign values to a and b to show that the rule $\dfrac{1}{a + b} = \dfrac{1}{a} + \dfrac{1}{b}$ is false.

SOLUTION

You can choose any values of a and b, except $a = -b$, $a = 0$, or $b = 0$. Let $a = 3$ and $b = 1$. Evaluate the left side of the equation.

$$\frac{1}{a + b} = \frac{1}{3 + 1} \qquad \text{Substitute 3 for } a \text{ and 1 for } b.$$

$$= \frac{1}{4} \qquad \text{Simplify.}$$

Lesson 12.8

Reteaching with Practice

For use with pages 758–764

Evaluate the right side of the equation.

$$\frac{1}{a} + \frac{1}{b} = \frac{1}{3} + \frac{1}{1} \qquad \text{Substitute 3 for } a \text{ and 1 for } b.$$

$$= \frac{4}{3} \qquad \text{Simplify.}$$

Because $\frac{1}{4} \neq \frac{4}{3}$, you have shown one case in which the rule in false.

The counterexample of $a = 3$ and $b = 1$ is sufficient to prove that
$\frac{1}{a + b} = \frac{1}{a} + \frac{1}{b}$ is false.

Exercises for Example 2

Find a counterexample to show that the statement is not true.

3. $\sqrt{a^2 + b^2} = a + b$ **4.** $a - b = b - a$ **5.** $a \div b = b \div a$

EXAMPLE 3 *Using an Indirect Proof*

Use an indirect proof to prove the conclusion is true:

If $\dfrac{a}{b} \geq \dfrac{c}{b}$ and $b > 0$, then $a \geq c$.

SOLUTION

Assume that $a \geq c$ is not true. Then $a < c$.

$a < c$ Assume the opposite of $a \geq c$ is true.

$\dfrac{a}{b} < \dfrac{c}{b}$ Dividing each side by the same positive number produces an equivalent inequality.

This contradicts the given statement that $\dfrac{a}{b} \geq \dfrac{c}{b}$. Therefore, it is impossible that $a < c$.

You conclude that $a \geq c$ is true.

Exercise for Example 3

6. Use an indirect proof to prove that the conclusion is true:
If $a + b > b + c$, then $a > c$.

NAME _____ DATE _____

Quick Catch-Up for Absent Students

For use with pages 758–764

The items checked below were covered in class on (date missed) _____

Lesson 12.8: Logical Reasoning: Proof

____ **Goal 1:** Use logical reasoning and proof to prove a statement is true. (pp. 758–759)

Material Covered:

____ Student Help: Study Tip

____ Example 1: Proving a Theorem

____ Example 2: Goldbach's Conjecture

Vocabulary:

postulate, p. 758   axiom, p. 758

theorem, p. 759   conjecture, p. 759

____ **Goal 2:** Prove that a statement is false. (pp. 760–761)

Material Covered:

____ Student Help: Look Back

____ Example 3: Finding a Counterexample

____ Example 4: Indirect Proof in Real Life

____ Example 5: Using an Indirect Proof

Vocabulary:

indirect proof, p. 760

____ Other (specify) _____

_____

Homework and Additional Learning Support

____ Textbook (specify) _pp. 761–764_____

_____

____ *Reteaching with Practice* worksheet (specify exercises)_____

____ *Personal Student Tutor* for Lesson 12.8

NAME _____   DATE _____

Interdisciplinary Application

For use with pages 758–764

Eratosthenes

ASTRONOMY   Eratosthenes (276?–196? B.C.) was a Greek mathematician, astronomer, geographer, and poet, who measured the circumference of the Earth with extraordinary accuracy by determining astronomically the difference in latitude between the Egyptian cities of Syene (now Aswan) and Alexandria.

Eratosthenes measured the altitude of the noontime sun at Alexandria at its maximum on June 21, the Summer solstice, and found a value of 7.12 degrees from a vertical line. On that date, the Sun is directly overhead at noontime at Syene, in southern Egypt. He assumed the two cities were on the same meridian. From this measurement and his knowledge of the distance between Syene and Alexandria, he was able to estimate the circumference of the Earth to be about 40,000 kilometers.

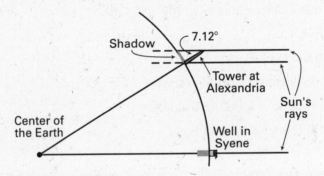

In Exercises 1 and 2, use the following information.

In geometry, there is a theorem that states: *If two parallel lines are cut by a transversal, then alternate interior angles are equal.*

1. Where in Eratosthenes's reasoning does he use this theorem?

2. Explain why this theorem is necessary to support his findings.

1. Prove the validity of the addition algorithm for $23 + 45$, that is, prove $23 + 45 = (20 + 40) + (3 + 5)$.

2. Prove the validity of the addition algorithm for any numbers of the form $10a + b$ and $10c + d$.

3. Prove the validity of the subtraction algorithm for any numbers of the form $10a + b$ and $10c + d$, that is, prove $(10a + b) - (10c + d) = (10a - 10c) + (b - d)$.

4. **a.** Find the product $23 \cdot 45$.

   b. Rewrite the multiplication from part (a) in the form $(10a + b)(10c + d)$ and use those values of a, b, c, and d to find $100ac + 10(bc + ad) + bd$. Do you get the same result as in part (a)?

   c. Prove the validity of the multiplication algorithm for any numbers of the form $10a + b$ and $10c + d$. (*Hint:* Prove $(10a + b)(10c + d) = 100ac + 10(bc + ad) + bd$.)

In Exercises 5–8, use the following information.

A *binary search* is an algorithm for locating an entry in an ordered list. For example, to find someone's name in the telephone directory, you could open to the middle page and narrow the search to one half of the directory by checking the first name on that page. Then you could open to the middle of the page *of that half* and repeat the process.

5. Suppose a classmate has thought of an integer between 1 and 64, inclusive, and you have to guess it by suggesting numbers and being told, "Too high," "Too low," or "Correct." Describe an algorithm for choosing your guesses based on a binary search.

6. What is the minimum number of guesses you need in order to be sure of guessing your classmate's number on or before the last guess?

7. Answer the question from Exercise 6 if your classmate's number is between 1 and 128; between 1 and 2^n.

8. Prove your minimum number of guesses is correct when your classmate chooses a number between 1 and 2^n.

NAME _____ DATE _____

Chapter Review Games and Activities

For use after Chapter 12

Solve the following problems in the space provided, and find the answer in the boxes at the bottom of the page. Cross out the box containing each correct answer. Place the remaining letters on the lines below the boxes to find the answer to the riddle.

Which trees are always misbehaving?

1. Simplify: $\sqrt{3}\left(2\sqrt{5} - 4\sqrt{8}\right)$

2. Simplify: $\left(1 - \sqrt{14}\right)^2$

3. Simplify: $\dfrac{18}{\sqrt{2}}$

4. Solve: $8 = \sqrt{3x - 2}$

5. Solve: $\sqrt{x - 7} + 6 = 10$

6. Find the missing side:

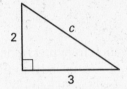

7. Find the missing side:

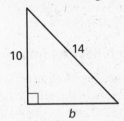

8. Find the distance between:
$(-7, 4)$ and $(8, -3)$

9. Find the distance between:
$(-11, 6)$ and $(-8, 12)$

23	$2\sqrt{15} - 4\sqrt{6}$	$4\sqrt{6}$	$3\sqrt{5}$	13
A	K	S	M	N
$15 + 2\sqrt{14}$	$2\sqrt{15} - 8\sqrt{6}$	$\sqrt{15}$	5	$\sqrt{13}$
O	A	T	T	R
$\sqrt{271}$	45	$15 - 2\sqrt{14}$	$9\sqrt{2}$	18
Y	P	T	E	I
$\sqrt{274}$	$6\sqrt{3}$	$3\sqrt{14} - 2\sqrt{6}$	22	$3\sqrt{6}$
P	N	E	R	S

___ ___ ___   ___ ___ ___   ___ ___ ___ ___ ___ !!

Identify the domain and range of the function. Then sketch its graph.

1. $y = 4\sqrt{x}$

2. $y = \sqrt{x} + 2$

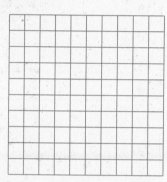

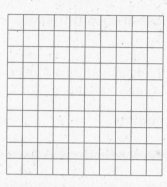

Simplify the expression.

3. $3\sqrt{5} + 4\sqrt{5}$

4. $6\sqrt{3} - 2\sqrt{3}$

5. $\sqrt{5} \cdot \sqrt{20}$

6. $\dfrac{3}{\sqrt{2}}$

7. $\sqrt{5}\left(6\sqrt{2} - \sqrt{5}\right)$

8. $\dfrac{4}{\sqrt{20}}$

Solve the equation.

9. $\sqrt{x} - 2 = 0$

10. $\sqrt{3x} - 6 = 0$

Two numbers and their geometric mean are given. Find the value of *a*.

11. 4 and a; 12

12. 3 and a; 9

Solve the equation by completing the square.

13. $x^2 + 2x = 3$

14. $x^2 + 8x = 14$

Find each missing length.

15.

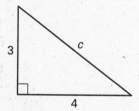

16.

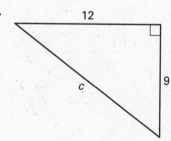

1.	_____
2.	_____
3.	_____
4.	_____
5.	_____
6.	_____
7.	_____
8.	_____
9.	_____
10.	_____
11.	_____
12.	_____
13.	_____
14.	_____
15.	_____
16.	_____

Review and Assess

Find the distance between the two points.

17. $(5, 8), (2, 4)$ **18.** $(7, 12), (1, 4)$

Decide whether the points are vertices of a right triangle.

19. $(0, 0), (0, 3), (4, 0)$ **20.** $(1, 2), (3, 0), (3, 3)$

21. You and a friend go biking. You bike 12 miles north and 2 miles east. What is the straight-line distance from your starting point?

Find the midpoint between the two points.

22. $(2, 2), (6, 4)$ **23.** $(2, 3), (4, 1)$

24. A company had sales of $500,000 in 1996 and sales of $720,000 in 1998. Use the midpoint formula to find the company's sales in 1997.

Find the sine, cosine, and tangent of $\angle R$ and of $\angle S$.

17.
18.
19.
20.
21.
22.
23.
24.
25.
26.
27.
28.

25.

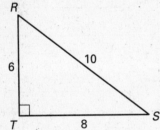

26.

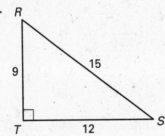

Find the missing lengths of the sides of the triangles. Round your answer to the nearest hundredth.

27.

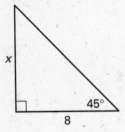

28.

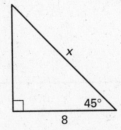

NAME _____ DATE _____

Chapter Test B

For use after Chapter 12

Identify the domain and range of the function. Then sketch its graph.

1. $y = \sqrt{x} - 2$

2. $y = \sqrt{x} + 1$

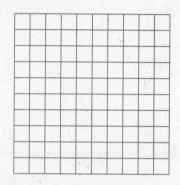

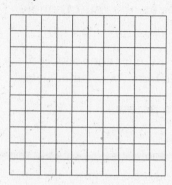

Simplify the expression.

3. $\sqrt{3} + \sqrt{27}$

4. $7\sqrt{5} - \sqrt{125}$

5. $\sqrt{3}(4\sqrt{2} + \sqrt{3})$

6. $\dfrac{7}{\sqrt{5}}$

7. $(2 + \sqrt{3})(2 - \sqrt{3})$

8. $\dfrac{6}{8 - \sqrt{3}}$

Solve the equation.

9. $\sqrt{x} - 3 = 0$

10. $\sqrt{4x} - 12 = 0$

Two numbers and their geometric mean are given. Find the value of a.

11. 2 and a; 8

12. 8 and a; 16

Solve the equation by completing the square.

13. $x^2 + 6x = 4$

14. $x^2 + x = 2$

Find each missing length.

15.

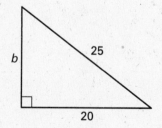

16.

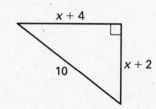

1.	_____
2.	_____
3.	_____
4.	_____
5.	_____
6.	_____
7.	_____
8.	_____
9.	_____
10.	_____
11.	_____
12.	_____
13.	_____
14.	_____
15.	_____
16.	_____

Review and Assess

Find the distance between the two points.

17. (6, 5), (2, 2)   **18.** (−2, 8), (3, 4)

Decide whether the points are vertices of a right triangle.

19. (1, 1), (1, 4), (5, 1)   **20.** (1, 2), (1, −1), (3, 0)

21. You and a friend go biking. You bike 13 miles north and 5 miles east. What is the straight-line distance from your starting point?

Find the midpoint between the two points.

22. (−1, 3), (4, 2)   **23.** (−4, 3), (3, 1)

24. A company had sales of $525,000 in 1996 and sales of $750,000 in 1998. Use the midpoint formula to find the company's sales in 1997.

Find the sine, cosine, and tangent of ∠R and of ∠S.

25.

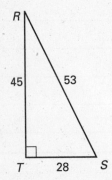

26.

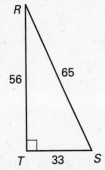

17. _____
18. _____
19. _____
20. _____
21. _____
22. _____
23. _____
24. _____
25. _____
_____
_____
26. _____
_____
_____
_____
27. _____
28. _____

Find the missing lengths of the sides of the triangles. Round your answer to the nearest hundredth.

27.

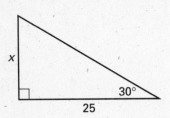

28.

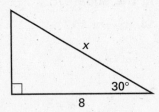

NAME _____ DATE _____

Chapter Test C

For use after Chapter 12

Identify the domain and range of the function. Then sketch its graph.

1. $y = \sqrt{x + 5}$

2. $y = \sqrt{2x + 3}$

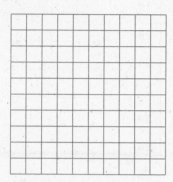

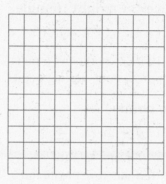

Simplify the expression.

3. $\sqrt{54} + \sqrt{24}$

4. $\sqrt{28} - \sqrt{175}$

5. $\left(5\sqrt{3} + 4\right)^2$

6. $\dfrac{8}{\sqrt{5} + 4}$

7. $7\sqrt{5} - \left(3\sqrt{5} + 6\right)^2$

8. $\dfrac{2 + \sqrt{7}}{2 - \sqrt{7}}$

Solve the equation.

9. $\sqrt{3x} - 3 = 0$

10. $\sqrt{2x + 1} - 5 = 0$

Two numbers and their geometric mean are given. Find the value of a.

11. 6 and a; 96

12. 4 and a; 108

Solve the equation by completing the square.

13. $x^2 - x = 5$

14. $2x^2 + 3x = 4$

Find each missing length.

15.

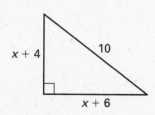

16.

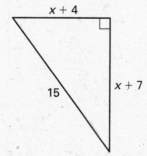

1. _____

2. _____

3. _____

4. _____

5. _____

6. _____

7. _____

8. _____

9. _____

10. _____

11. _____

12. _____

13. _____

14. _____

15. _____

16. _____

Review and Assess

NAME _____ DATE _____

Chapter Test C

For use after Chapter 12

Find the distance between the two points.

17. $(2, -2), (-1, -2)$ **18.** $(-4, -3), (2, 5)$

Decide whether the points are vertices of a right triangle.

19. $(-1, -1), (-3, -1), (-3, 3)$ **20.** $(-3, 2), (-1, -1), (-4, -2)$

21. You and a friend go biking. You bike 10 miles north and 6 miles west. What is the straight-line distance from your starting point?

Find the midpoint between the two points.

22. $(-2, 4), (3, -5)$ **23.** $(-5, -2), (3, -4)$

24. A company had sales of $2,310,000 in 1996 and sales of $4,515,000 in 1998. Use the midpoint formula to estimate the company's sales in 1997.

Find the sine, cosine, and tangent of $\angle R$ and of $\angle S$.

25.

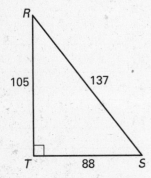

26.

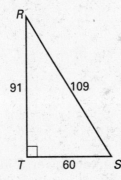

Find the missing lengths of the sides of the triangles. Round your answers to the nearest hundredth.

27.

28.

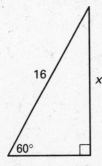

17. _____

18. _____

19. _____

20. _____

21. _____

22. _____

23. _____

24. _____

25. _____

_____

_____

_____

26. _____

_____

_____

_____

27. _____

28. _____

1. What are the domain and range of
   $y = \sqrt{x + 1} - 2$?

   A domain: all nonnegative numbers
   range: all nonnegative numbers

   B domain: $x \geq -1$
   range: all numbers greater than or equal to -2

   C domain: all nonnegative numbers
   range: all numbers greater than or equal to -2

   D domain: $x \geq -1$
   range: all nonnegative numbers

2. Which of the following is the difference
   $\sqrt{112} - \sqrt{63}$, in simplified form?

   A $\sqrt{112} - 3\sqrt{7}$ B $\sqrt{112} - 9\sqrt{7}$

   C $-5\sqrt{7}$ D $\sqrt{7}$

3. Simplify $\dfrac{7}{6 - 4\sqrt{3}}$.

   A $\dfrac{-42 + 11\sqrt{3}}{12}$ B $\dfrac{60 + \sqrt{3}}{-12}$

   C $\dfrac{-21 - 14\sqrt{3}}{6}$ D $\dfrac{-21 + 14\sqrt{3}}{54}$

4. Solve $\sqrt{3x - 5} - 2 = 0$.

   A $x = \dfrac{1}{3}$ B $x = 3$

   C $x = \dfrac{49}{3}$ D $x = 7$

5. What is the distance between $(-2, 3)$ and $(3, -4)$?

   A $\sqrt{2}$ B $2\sqrt{2}$

   C $\sqrt{26}$ D $\sqrt{74}$

In Questions 6 and 7, choose the statement below that is true about the given numbers.

   A. The number in column A is greater.

   B. The number in column B is greater.

   C. The two numbers are equal.

   D. The relationship cannot be determined from the given information.

6.

Column A	Column B
geometric mean of 5 and 35	geometric mean of 3 and 49

   A   B   C   D

7.

Column A	Column B
The solution of $\sqrt{x} - 4 = 3$	The solution of $\sqrt{x - 4} = 3$

   A   B   C   D

8. What are the sine, cosine, and tangent of $\angle R$?

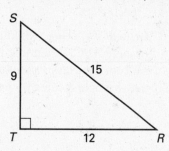

   A sine: $\frac{3}{5}$, cosine: $\frac{4}{5}$, tangent: $\frac{3}{4}$

   B sine: $\frac{4}{5}$, cosine: $\frac{3}{5}$, tangent: $\frac{3}{4}$

   C sine: $\frac{4}{5}$, cosine: $\frac{3}{5}$, tangent: $\frac{4}{3}$

   D sine: $\frac{3}{5}$, cosine: $\frac{4}{5}$, tangent: $\frac{4}{3}$

Review and Assess

Alternative Assessment and Math Journal

For use after Chapter 12

JOURNAL 1. Consider the graphs of $y = |x|$, $y = x^2$, and $y = \sqrt{x}$ to be the "parent" functions. (a) Explain what the graphs of $y = |x| + 2$, $y = x^2 + 2$, and $y = \sqrt{x} + 2$ have in common. Describe how the graphs of $y = |x| - 1$, $y = x^2 - 1$, and $y = \sqrt{x} - 1$ differ from their parent functions. Describe the graphs of $y = |x + 3|$, $y = (x + 3)^2$, and $y = \sqrt{x + 3}$ compared to the graphs of the parent functions. (b) Given a generic equation $y = |x - h| + k$, $y = (x - h)^2 + k$, or $y = \sqrt{x - h} + k$, what effect do h and k have on the graphs? Write a paragraph to explain different possible scenarios. Be sure to consider values that are positive, negative, or zero. (c) How would a negative sign in front of a function (such as $y = -\sqrt{x - 2}$) change the graph?

MULTI-STEP 2. Consider the triangle with vertices $A(2, -3)$, $B(8, 3)$, and $C(8, -9)$ to
PROBLEM   answer the following. Graph and label this triangle.

   a. Find the midpoint of each side of the triangle.

   b. Graph the new triangle formed by connecting the midpoints found in part (a). Find the length of each side of this new triangle. Give exact answers.

   c. Determine if the triangle formed in part (b) is a right triangle. Explain how you know.

   d. Find the perimeter and area of the triangle from part (b).

   e. Find and simplify $\sin B$, $\cos B$, and $\tan B$. Give exact answers.

   f. Given point $D(-4, y)$ and the distance from point A to D is 10 units, find the possible value(s) for y. Show all work and give reasons for each step.

   3. *Critical Thinking* Find the lengths of all three sides of the original triangle ABC. Determine if the original triangle ABC is a right triangle. Make a conjecture about the relationship between a triangle and the new triangle formed by joining the midpoints.

Alternative Assessment Rubric

For use after Chapter 12

JOURNAL SOLUTION

1. **a–c.** Complete answers should address these points.

    a. • Explain that the graphs all shift up 2 units when 2 is added to the function, and explain how they all shift down when 1 is subtracted from the function.

    • Explain that when 3 is added to the *x*-value in the function, the graphs are shifted 3 units to the left from the parent function.

    b. • Explain that if *h* is positive, the graph will shift to the right *h* units, and if *h* is negative, the graph will shift to the left $|h|$ units. If *k* is positive, the graph will shift up *k* units, and if *k* is negative, the graph will shift down $|k|$ units from the parent graph. If either *h* or *k* is zero, the graph does not shift for that portion.

    c. • Explain that a negative sign in front of the function will cause a reflection of the graph in the *x*-axis.

MULTI-STEP PROBLEM SOLUTION

2. **a.** $(5, 0), (8, -3), (5, -6)$

    b. $3\sqrt{2}, 3\sqrt{2},$ and 6

    c. Yes, it is a right triangle because the sides fit the conditions of the Pythagorean theorem.

    d. $6 + 6\sqrt{2}$ units; 9 square units

    e. $\sin B = \cos B = \dfrac{\sqrt{2}}{2}; \tan B = 1$

    f. $5, -11$

3. $6\sqrt{2}, 6\sqrt{2}, 12;$ The original triangle is a right triangle. *Sample answer:* Since the original triangle is a right triangle, the new triangle formed by connecting its midpoints must also be a right triangle because the triangles are similar triangles.

MULTI-STEP PROBLEM RUBRIC

4 Students complete all parts of the questions accurately. Explanations are logical and clear. Students are able to apply the formulas from the chapter correctly. Exact and simplified answers are given when required. Students correctly use the trigonometric ratios.

3 Students complete questions and explanations. Solutions may contain minor mathematical errors or misunderstandings. Students are able to apply the formulas from the chapter. Exact and simplified answers are given when required. Students use the trigonometric ratios.

2 Students complete questions and explanations. Several mathematical errors may occur. Students are not always able to apply the formulas from the chapter correctly. Exact or simplified answers are not given. Students do not correctly use the trigonometric ratios.

1 Students' work is very incomplete. Solutions and reasoning are incorrect. Students are not able to apply the formulas from the chapter. Exact and simplified answers are not given. Students do not use the trigonometric ratios correctly.

Project: Puzzling Distances

For use with Chapter 12

OBJECTIVE **Create a puzzle using the midpoint and distance formulas.**

MATERIALS graph paper, pencil

INVESTIGATION Draw a set of axes on a piece of graph paper. Draw a design on the graph paper using between 10 and 30 straight line segments with endpoints that have integer coordinates. Find the coordinates of the endpoints of your line segments. Make a puzzle using the midpoint and distance formulas so that your design is the solution to the puzzle. Solve the following part of a puzzle to help you get started.

1. Find the point (a, b) so that (a, b) is 5 units from the origin, $a = b + 1$, $a > 0$, and $b > 0$.

2. Connect (a, b) to the midpoint of the segment between (a, b) and $(0, b)$.

3. Find the point $(0, c)$ so the distance between $(0, c)$ and (a, b) is $2\sqrt{5}$ and $c > 2$. Connect the midpoint from Step 2 to $(0, c)$.

4. Connect (a, b) and $(a, b - 2)$. Then connect $(a, b - 2)$ and $(c, 0)$.

You may want to use transformations in your design and in your instructions, in addition to steps involving the distance and midpoint formulas.

PRESENT YOUR RESULTS Compile your puzzle with those of your classmates into a book with answers at the end.

Review and Assess

GOALS
- Solve a quadratic equation and a radical equation.
- Find the distance and midpoint between two points in a coordinate plane.
- Translate between one mathematical representation and another.

MANAGING THE PROJECT
You may wish to have students work individually to create their own puzzles, but then have them trade puzzles with a partner to solve. Partners can give each other suggestions for clarifying directions or otherwise improving the puzzle. Encourage students to include directions that require the solving of a radical equation.

You might also want to suggest that students use special right triangles in their designs to make their calculations easier.

RUBRIC
The following rubric can be used to assess student work.

4 The student develops a design, gives clear and accurate directions involving an appropriate degree of difficulty, and provides the correct solution. The design is interesting and includes a degree of complexity.

3 The student develops a design, gives accurate directions, and provides the solution. However, the design may be somewhat simplistic, the directions may not be as clear as possible or may not have an appropriate degree of difficulty, or there may be errors in the directions or the solution.

2 The student develops a design, gives directions, and provides a solution. However, work may be incomplete or reflect misunderstanding. For example, the student may not use both the distance and the midpoint formula in the directions or the directions may not match the solution provided due to errors in computation. The design may be simplistic.

1 Part of the design, directions, or solution are missing or do not show an understanding of key ideas.

Cumulative Review

For use after Chapters 1–12

Evaluate the expression. (1.3)

1. $12 \div (10 - 7)^3 + \frac{1}{4}$

2. $y \div \frac{1}{2} - y + y^2$ when $y = -2$

3. $\dfrac{r + 19}{2} - \dfrac{r}{s} + r^2$ when $r = -3$, $s = -2$

4. $xy \div \dfrac{1}{x} - y^2 + 10$ when $x = -\dfrac{1}{3}$, $y = \dfrac{8}{3}$

The two triangles are similar. Find the length of the missing side. (3.2)

5.
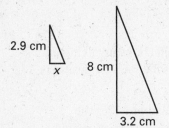

6.
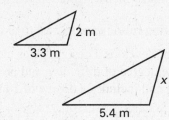

Decide whether the graphs of the two equations are parallel. (4.6)

7. $y = -7x + 15, 7y = x - 15$

8. $4x - y = 12, \frac{1}{4}x + y = 9$

9. $-x + y = 12, y = x - 20$

10. $\frac{2}{3}x = \frac{1}{7}y - \frac{3}{7}, y = -\frac{3}{4}x - 3$

Sketch the graph of the inequality. (6.5)

11. $x + 2 \leq 5y$

12. $7x + 9y \geq 16$

13. $\frac{1}{5}x > \frac{2}{7}y - 2$

You buy a used boat for $15,000. It depreciates at the rate of 15% per year. Find the value of the boat in the given number of years. (8.6)

14. 2 years

15. 9 years

16. 15 years

Simplify the expression. (9.2)

17. $\sqrt{\dfrac{5}{12}}$

18. $\dfrac{\sqrt{64}}{\sqrt{8}}$

19. $\sqrt{\dfrac{800}{245}}$

Factor the expression completely. (10.8)

20. $-96t^3 - 84t^2 + 12t$

21. $3t^6 - 21t^5 - 24t^4$

Simplify the expression. (11.5)

22. $\dfrac{x + 9}{x^2 + 8x - 9} \div \dfrac{1}{x^3 - x^2}$

23. $\dfrac{x^2}{x^2 - 2x} \cdot \dfrac{x^4 - 4x^2}{x} \cdot \dfrac{x^3}{x^2 + 4x + 4}$

Cumulative Review

For use after Chapters 1–12

Find the domain of the function. (12.1)

24. $y = \sqrt{x - 4}$

25. $y = 6\sqrt{x}$

26. $x = \sqrt{8 + x}$

27. $y = \sqrt{9x - 1}$

28. $y = x\sqrt{x + \dfrac{1}{2}}$

29. $y = \dfrac{\sqrt{1 - x}}{x}$

Simplify the expression. (12.2)

30. $\sqrt{8} + \sqrt{162} + \sqrt{288}$

31. $\sqrt{245} - \sqrt{320} + \sqrt{500}$

32. $\left(4 - \sqrt{19}\right)\left(4 + \sqrt{19}\right)$

33. $7\sqrt{6} + \left(2 - \sqrt{6}\right)^2$

Solve the equation. (12.3)

34. $\sqrt{x} - 34 = 0$

35. $\sqrt{4x + 1} + 3 = 10$

36. $\sqrt{24 - 5x} - x = 0$

37. $x = \sqrt{x + 2}$

Solve the equation by completing the square. (12.4)

38. $x^2 - 4x + 2 = 0$

39. $-x^2 - 5x = -8$

40. $2x^2 = \dfrac{4}{3}x + 6$

41. $x^2 - 1 = -\dfrac{4}{5}x$

Find the missing lengths. (12.5)

42.

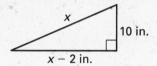

43.

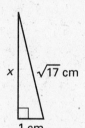

44.

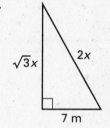

Find the distance between the two points. Round the result to the nearest hundredth if necessary. (12.6)

45. $(3, 9), (1, -2)$

46. $(-1, 8), (-7, 9)$

47. $(2, 9), (-2, -9)$

48. $\left(\dfrac{1}{3}, \dfrac{1}{6}\right), (2, 0)$

49. $(-0.1, 0.2), (3, -6)$

50. $(0, -8), (-2.02, 8.98)$

Find the sine, the cosine, and the tangent of $\angle R$. (12.7)

51.

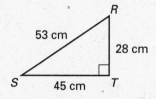

52.

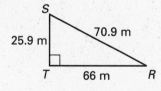

53.

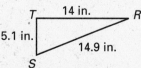

Review and Assess

ANSWERS

Chapter Support

Parent Guide
Chapter 12

12.1: about 3 cm   **12.2:** $2\sqrt{3} + \sqrt{6}$

12.3: 2000 m   **12.4:** $1 \pm \sqrt{5}$ **12.5:** The gable has 2 right triangles, each 9 ft by 12 ft by 15 ft and $9^2 + 12^2 = 15^2$ **12.6:** $(-1, -1)$; 10

12.7: about 34.6 ft

12.8: *Sample answer:* $\frac{1}{2}(6) < 6$

Prerequisite Skills Review

1. 6 cm   **2.** 1.35 cm   **3.** $\frac{15}{19}$ in.   **4.** 14 in.

5. $\dfrac{\sqrt{5}}{12}$ **6.** $\dfrac{\sqrt{6}}{2}$ **7.** $\dfrac{2\sqrt{3}}{5}$ **8.** $\dfrac{\sqrt{5}}{5}$

9. $(x - 2)(x + 10)$ **10.** $(x + 4)(x + 16)$

11. $(x - 8)(x - 9)$

Strategies for Reading Mathematics

1. Yes; triangle *ABD* and triangle *CBD*.

2. All three sides of triangle *ABC* are the same length; *D* is the midpoint of $\overline{AC}$.

3. If one angle of the triangle contains the right angle symbol, then the Pythagorean Theorem applies to the triangle.

4. Check diagrams. All four angles should contain the right angle symbol; the four sides should all be labeled with the same length, say *x*.

Lesson 12.1

Warm-up Exercises

1. 1.7   **2.** 2   **3.** 5   **4.** *a*

Daily Homework Quiz

1. $-\frac{21}{2}$ **2.** -4 **3.** $-\frac{5}{2}, 4$ **4.** $\frac{1}{3}$

---

5. 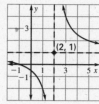   domain: all real numbers except 2

6. 4

Lesson Opener
Allow 10 minutes.

1. C; the square root of the area of a square is equal to the length of its side.

2. C; the square root of the area of a square is equal to the length of its side.

3. a. 3   **b.** 4   **c.** 5   **d.** 6   **e.** 7   **f.** 10

4. find the positive square root of the area

Practice A

1. 3 and 4   **2.** 6 and 7   **3.** 9 and 10

4. 5 and 6   **5.** 8 and 9   **6.** 7 and 8

7. 11 and 12   **8.** 10 and 11   **9.** 20   **10.** $\frac{3}{2}$

11. 4.8   **12.** 4.4   **13.** 4.9   **14.** 7.9   **15.** $x \geq 0$

16. $x \geq 4$ **17.** $x \geq -8$ **18.** $x \geq -2$

19. $x \geq 0$ **20.** $x \geq -\frac{1}{2}$ **21.** $x \geq -1$

22. $x \geq -3$ **23.** $x \geq -\frac{3}{5}$ **24.** B   **25.** C

26. A   **27.** 1.59 m^2

Practice B

1. 18   **2.** 1   **3.** 5.4   **4.** 3.9   **5.** 7.1   **6.** 17.0

7. 5   **8.** 7.5   **9.** 31.6   **10.** $x \geq 0$ **11.** $x \geq 9$

12. $x \geq -4.5$ **13.** $x \geq 0$ **14.** $x \geq 7$

15. $x \geq -6$ **16.** $x \geq -\frac{1}{2}$ **17.** $x \geq 0$

18. $x \geq \frac{4}{3}$ **19.** $x \geq 0$ **20.** $x \geq \frac{2}{7}$ **21.** $x \geq \frac{1}{3}$

22. B   **23.** C   **24.** A

25. domain: $x \geq 0$
range: $y \geq 0$

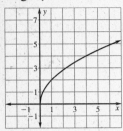

26. domain: $x \geq 0$
range: $y \geq 6$

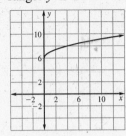

25. domain: $x \geq 0$
range: $y \geq 0$

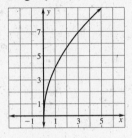

26. domain: $x \geq 0$
range: $y \geq 9$

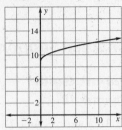

27. domain: $x \geq 0$
range: $y \geq -5$

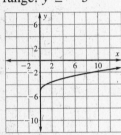

28. domain: $x \geq -\frac{1}{2}$
range: $y \geq 0$

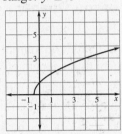

27. domain: $x \geq 0$
range: $y \geq -2.5$

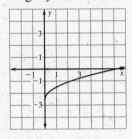

28. domain: $x \geq -\frac{2}{5}$
range: $y \geq 0$

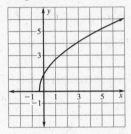

29. domain: $x \geq -6$
range: $y \geq 0$

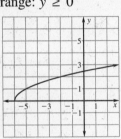

30. domain: $x \geq 4$
range: $y \geq 0$

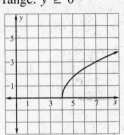

29. domain: $x \geq \frac{1}{3}$
range: $y \geq 0$

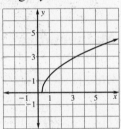

30. domain: $x \geq \frac{1}{3}$
range: $y \geq 0$

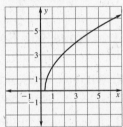

31. $r \approx 4.15$ in.

Practice C

 1. 56 **2.** $\frac{5}{3}$ **3.** 8.4 **4.** 4.9 **5.** 8.9 **6.** 50.9

 7. 3.7 **8.** 6.9 **9.** 30.2 **10.** 24.7 **11.** 6.7

12. 6.2 **13.** $x \geq 0$ **14.** $x \geq 3$ **15.** $x \geq -6.5$

16. $x \geq -\frac{3}{4}$ **17.** $x \geq 0$ **18.** $x \geq \frac{2}{5}$ **19.** $x \geq 0$

20. $x \geq \frac{4}{9}$ **21.** $x \geq \frac{1}{6}$ **22.** $x \geq -10$ **23.** $x \geq 2$

24. all real numbers

31. domain: $x \geq 0$
range: $y \geq 3$

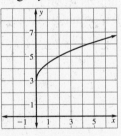

32. domain: $x \geq -2$
range: $y \geq -3$

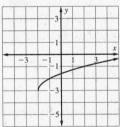

33. domain: $x \geq 5$; range: $y \geq 1$

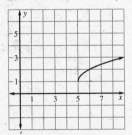

Lesson 12.1 *continued*

34.

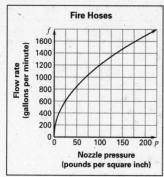

Fire Hoses

35. 44.4 lbs/in.^{2}

36. No; the nozzle pressure quadruples to about 177.8 lb/sq in.

Reteaching with Practice

1. The domain is the set of all nonnegative numbers. The range is the set of all numbers greater than or equal to 1.

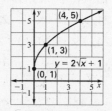

2. The domain is the set of all nonnegative numbers. The range is the set of all numbers greater than or equal to −1.

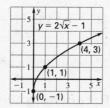

3. The domain is the set of all nonnegative numbers. The range is the set of all numbers greater than or equal to −2.

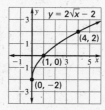

4. The domain is the set of all numbers greater than or equal to 1. The range is the set of all nonnegative numbers.

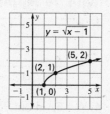

5. The domain is the set of all numbers greater than or equal to −1. The range is the set of all nonnegative numbers.

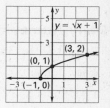

6. The domain is the set of all numbers greater than or equal to 4. The range is the set of all nonnegative numbers.

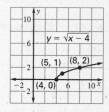

7. 48 ft/sec

Interdisciplinary Application

1. $a^2 + b^2 = r^2$ **2.** $r = \sqrt{a^2 + b^2}$

3. $\sqrt{2}$ **4.** $\sqrt{3}$

5.

n	0	1	2	3	4	5	6
r_n	$\sqrt{1}$	$\sqrt{2}$	$\sqrt{3}$	$\sqrt{4}$	$\sqrt{5}$	$\sqrt{6}$	$\sqrt{7}$

n	7	8	9	10	11
r_n	$\sqrt{8}$	$\sqrt{9}$	$\sqrt{10}$	$\sqrt{11}$	$\sqrt{12}$

n	12	13	14	15
r_n	$\sqrt{13}$	$\sqrt{14}$	$\sqrt{15}$	$\sqrt{16}$

6. $A_n = \frac{1}{2}(r_{n-1})$ **7.** $\sqrt{2}$

Challenge: Skills and Applications

1. all numbers $x > -9$ **2.** all numbers $x < -\frac{8}{3}$

3. all numbers $x \le -2$ or $x \ge 7$

4. all numbers $x \le 0$ or $x \ge 9$

5. all numbers $0 \le x \le 6$

6. all numbers $3 \le x \le 5$ **7.** about 8.8%

Lesson 12.2

Warm-up Exercises

1. $-3a$ **2.** $x^2 - 9$ **3.** $4x^2 - 81$

4. $\dfrac{5 \pm \sqrt{13}}{2}$

Answers

Lesson 12.2 *continued*

Daily Homework Quiz

1. a. 3.7 **b.** 38.2

2. a. domain: $x \geq \frac{7}{2}$, range: $y \geq 0$

b. domain: $x \geq 0$, range: $y \geq 7$

c. domain: $x \leq 3$, $x \neq 0$, range: all reals

Lesson Opener

Allow 10 minutes.

1. $5x; 5\sqrt{2}$ **2.** $7y; 7\sqrt{3}$ **3.** $4x; 4\sqrt{5}$

4. $6z; 6\sqrt{10}$ **5.** $6a; 6\sqrt{7}$

6. $x^2; \left(\sqrt{2}\right)^2 = \sqrt{4} = 2$

7. $2y^2; 2\left(\sqrt{5}\right)^2 = 2\sqrt{25} = 10$

8. $24z^2; 24\left(\sqrt{3}\right)^2 = 24\sqrt{9} = 72$

Practice A

1. $4\sqrt{2}$ **2.** $5\sqrt{3}$ **3.** $3\sqrt{3}$ **4.** $6\sqrt{2}$

5. $2\sqrt{5}$ **6.** $2\sqrt{3}$ **7.** $5\sqrt{5}$ **8.** $4\sqrt{3}$

9. $2\sqrt{3}$ **10.** $3\sqrt{5}$ **11.** $4\sqrt{5}$ **12.** $7\sqrt{3}$

13. $15\sqrt{6}$ **14.** $-\sqrt{8}$ **15.** $6\sqrt{2}$ **16.** 0

17. $2\sqrt{2}$ **18.** $6\sqrt{5}$ **19.** $-\sqrt{7}$ **20.** $13\sqrt{6}$

21. 4 **22.** 6 **23.** $4\sqrt{2}$ **24.** $6\sqrt{2}$

25. $3\sqrt{10} + 2\sqrt{5}$ **26.** -12 **27.** -6

28. $9 + 4\sqrt{5}$ **29.** $\frac{1}{2}\sqrt{6}$ **30.** $\frac{3}{2}\sqrt{2}$ **31.** $\frac{7}{3}\sqrt{3}$

32. $8 - 4\sqrt{3}$ **33.** yes **34.** yes **35.** yes

36. yes **37.** 32 ft/sec **38.** $4\sqrt{3} + 14\sqrt{2}$

39. $3\sqrt{5} - \sqrt{10}$ **40.** $3\sqrt{2} + 4\sqrt{3}$

Practice B

1. $7\sqrt{2}$ **2.** $4\sqrt{3}$ **3.** $3\sqrt{6}$ **4.** $9\sqrt{2}$

5. $8\sqrt{3}$ **6.** $4\sqrt{7}$ **7.** $5\sqrt{10}$ **8.** $6\sqrt{3}$

9. $6\sqrt{3}$ **10.** $9\sqrt{5}$ **11.** $4\sqrt{2}$ **12.** $-3\sqrt{6}$

13. $3\sqrt{5}$ **14.** $-2\sqrt{3}$ **15.** $7\sqrt{2}$ **16.** $-\sqrt{3}$

17. $3\sqrt{2}$ **18.** $2\sqrt{7}$ **19.** $-2\sqrt{2}$ **20.** $-5\sqrt{10}$

21. 4 **22.** 6 **23.** $3\sqrt{7}$ **24.** $6 + \sqrt{10}$

25. $\sqrt{35}$ **26.** $6\sqrt{2} - 8\sqrt{3}$ **27.** $7 + 4\sqrt{3}$

28. $21 - 8\sqrt{5}$ **29.** $19 - 6\sqrt{2}$ **30.** $2\sqrt{3}$

31. $\sqrt{5}$ **32.** $\frac{7\sqrt{2}}{2}$ **33.** $\frac{12 - 3\sqrt{2}}{14}$

34. $-2 - \sqrt{3}$ **35.** $\frac{10 - 2\sqrt{2}}{23}$ **36.** yes

37. no **38.** no **39.** yes

40. about 7071;

t	0	1	2	3	4	5
N	7071	7294	7510	7720	7925	8124

t	6	7	8	9
N	8319	8509	8695	8877

41. $21 + 24\sqrt{7}; 16 + 8\sqrt{7}$

Practice C

1. $-4\sqrt{3}$ **2.** $15\sqrt{3}$ **3.** $11\sqrt{6}$ **4.** $-2\sqrt{2}$

5. $25\sqrt{5}$ **6.** $\frac{1}{4}\sqrt{2}$ **7.** $8\sqrt{3}$

8. $-4\sqrt{10} + 4\sqrt{5}$ **9.** $4\sqrt{10} - \sqrt{5}$

10. $9\sqrt{2}$ **11.** $17\sqrt{2}$ **12.** $19\sqrt{3}$

13. $-2\sqrt{5} - 6\sqrt{6}$ **14.** $11\sqrt{5}$

15. $9\sqrt{10} + 60$ **16.** $2\sqrt{21}$ **17.** $4\sqrt{10}$

18. $8\sqrt{6}$ **19.** $6 + 4\sqrt{6}$ **20.** $35 - 5\sqrt{2}$

21. $12 - 6\sqrt{6}$ **22.** -1 **23.** 11

24. $79 - 20\sqrt{3}$ **25.** $\frac{4}{7}\sqrt{7}$ **26.** $\frac{5}{4}\sqrt{6}$

27. $-4\sqrt{6}$ **28.** $\frac{16 - 2\sqrt{11}}{53}$ **29.** $\frac{6\sqrt{5} + 9}{11}$

30. $\frac{18 - 3\sqrt{5}}{31}$ **31.** yes **32.** no **33.** no

34. yes **35.** $15 + 12\sqrt{5}; 8\sqrt{5} + 8$

36. $\frac{3}{2}\sqrt{3}; 3\sqrt{3} + 3$ **37.** $3; 2\sqrt{6} + 2\sqrt{3}$

38. about 90.8 volts

Reteaching with Practice

1. $4\sqrt{7}$ **2.** $\sqrt{2}$ **3.** $5\sqrt{3}$ **4.** $3 + 2\sqrt{2}$

5. $3\sqrt{2}$ **6.** $2\sqrt{10} + 2\sqrt{5}$ **7.** $\frac{4\sqrt{3}}{3}$

8. $\frac{5\sqrt{2}}{4}$ **9.** $-\frac{\sqrt{3}}{6}$ **10.** $19.3\sqrt{5}$ mph

Real-Life Application

1.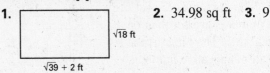

2. 34.98 sq ft **3.** 9

4. 5 **5.** about 75.4 feet **6.** 349.8 sq ft

Challenge: Skills and Applications

1. $\frac{17\sqrt{3}}{6}$ **2.** 0 **3.** $15x\sqrt{14} - 24$ **4.** $-54x$

Lesson 12.2 *continued*

5. $12\sqrt{6}$ **6.** $-35x$ **7.** $6x^2 - 3x\sqrt{2} - 6$

8. a. $-7 + 24i$ **b.** $67 - 36i$ **c.** 29

d. $a^2 + b^2$ **9.** $3\sqrt{x} + \sqrt{3x}$

10. $2\sqrt{x} + 2\sqrt{x - 2} + 2\sqrt{x - 3}$

11. $\sqrt{x} + \sqrt{3x} - 2\sqrt{x - 2} - 2\sqrt{x - 3}$

12. $\frac{1}{2}x\sqrt{3}$ **13.** $2\sqrt{x^2 - 5x + 6}$ **14.** 5

Lesson 12.3

Warm-up Exercises

1. 6 **2.** 16 **3.** 289 **4.** $x = 2$ or $x = 3$

5. $x = 1.8$

Daily Homework Quiz

1. a. $-4\sqrt{6}$ **b.** $4\sqrt{17} + 9\sqrt{11}$ **c.** $\sqrt{21}$

d. $7\sqrt{6} + 2$ **e.** $19 + 8\sqrt{3}$ **f.** $\dfrac{40 + 5\sqrt{6}}{58}$

2. $4 \pm \sqrt{17}$

Lesson Opener

Allow 10 minutes.

1. x; 16 **2.** x; 100 **3.** x; 49 **4.** 6; x; 36

5. $\sqrt{x}$; x; 25 **6.** $\sqrt{x}$; 3; x; 9

7. Square both sides.

Practice A

1. $\sqrt{x + 2} = 12$ **2.** $\sqrt{x} = 16$

3. $\sqrt{x - 7} = 2$ **4.** $\sqrt{5x - 4} = 2$

5. $\sqrt{2x - 1} = 0$ **6.** $\sqrt{x} = 4$

7. $\sqrt{\frac{1}{2}x - 4} = 12$ **8.** $\sqrt{4 - x} = \frac{1}{2}$

9. $\sqrt{x} = 6$ **10.** 25 **11.** 49 **12.** 81 **13.** 36

14. no solution **15.** 0 **16.** 1 **17.** 87 **18.** 66

19. no solution **20.** 3 **21.** 2 **22.** no solution

23. 7 **24.** $\sqrt{7}$ **25.** 6 **26.** 0 **27.** 1, 4

28. 18 **29.** 28 **30.** 45 **31.** 3.5 mi/hr

32. 1 in. **33.** 56

Practice B

1. 25 **2.** no solution **3.** 5 **4.** 32 **5.** -27

6. no solution **7.** 186 **8.** 6 **9.** $\frac{7}{4}$ **10.** 24

11. 180 **12.** $\frac{3}{25}$ **13.** no solution **14.** $-\frac{39}{10}$

15. $-\frac{3}{7}$ **16.** 3 **17.** 3 **18.** 2 **19.** no solution

20. 3 **21.** 2, 4 **22.** 7, 2 **23.** $\frac{5}{2}, \frac{3}{2}$

24. no solution **25.** $\frac{5}{8}$ **26.** 4 **27.** 4 **28.** 32

29. 75 **30.** 4 **31.** 2,415,000 **32.** about 14 ft

33. $48\pi \approx 151$ in.^{2} **34.** no; the area is $\frac{1}{4}$ the area of the fan with the 12 in. radius

Practice C

1. 100 **2.** 144 **3.** -4 **4.** 9 **5.** 45 **6.** 7

7. $\frac{100}{3}$ **8.** $-\frac{1}{3}$ **9.** -32 **10.** $\frac{1}{3}$

11. no solution **12.** 298 **13.** 40

14. no solution **15.** $\frac{18}{7}$ **16.** 4 **17.** 2

18. no solution **19.** 5 **20.** 7 **21.** no solution

22. 6 **23.** 0 **24.** 6 **25.** 5

26. $\frac{1}{3}$ **27.** $\frac{5}{2}$ **28.** $-\frac{2}{3}$ **29.** 2 **30.** 1, 3

31. no solution **32.** -4 **33.** 8 **34.** 16°C

35. 0 m/sec **36.** 7.30 ft

Reteaching with Practice

1. 7 **2.** 1 **3.** 2 **4.** 5 **5.** no solution **6.** 3

7. 2.66 cm **8.** 507 sec

Real-Life Application

1. ≈ 98.1 grams **2.** ≈ 5.4 ounces

3. right side: 34 grams; left side: 31 grams

Challenge: Skills and Applications

1. 18 **2.** 7 **3.** no solution **4.** no solution

5. 16 **6.** no solution **7.** $-3\sqrt{2}, 3\sqrt{2}$

8. $-2\sqrt{3}, 2\sqrt{3}$ **9.** $-3, 3$ **10.** decreases

11. about 1.8 m **12.** 2.1 m

Quiz 1

1. -1 **2.** 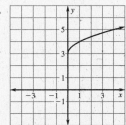 **3.** $4\sqrt{3}$

4. 4 **5.** 14 **6.** 45

Lesson 12.4

Warm-up Exercises

1. $x^2 + 18x + 81$ 2. $4x^2 - 12x + 9$
3. $(x - 5)^2$ 4. $(3x + 4)^2$

Daily Homework Quiz

1. **a.** 25 **b.** $\frac{107}{2}$ **c.** 14, 1 **2.** 27 **3.** 37

Lesson Opener

Allow 10 minutes.

1. $x^2 + 2x$ 2. $x^2 + 2x + 1 = 36$
3. $(x + 1)^2$; find the square root of each side, set them equal, and solve for x. 4. $x^2 - 4x$
5. $x^2 - 4x + 4 = 484$ 6. $(x - 2)^2$; find the square root of each side, set them equal, and solve for x. 7. $x^2 + 6x$ 8. $x^2 + 6x + 9 = 225$
9. $(x + 3)^2$; find the square root of each side, set them equal, and solve for x.

Practice A

1. 1, 6, 9 2. 1, -8, 16 3. 1, -12, 36
4. 1, 10, 25 5. 4, 12, 9 6. 25, -10, 1
7. 4 8. 49 9. 16 10. 1 11. 25 12. 64
13. $\frac{49}{4}$ 14. $\frac{9}{4}$ 15. $\frac{1}{16}$ 16. 0.16 17. 100
18. $\frac{1}{9}$ 19. 1, -7 20. 1, -5 21. 9, -1
22. 2, 10 23. $2 + \sqrt{2}, 2 - \sqrt{2}$
24. $-2 + \sqrt{5}, -2 - \sqrt{5}$
25. $-3 + \sqrt{13}, -3 - \sqrt{13}$
26. $1 + \sqrt{6}, 1 - \sqrt{6}$
27. $\frac{-3 + \sqrt{17}}{2}, \frac{-3 - \sqrt{17}}{2}$
28. $\frac{-1 + \sqrt{5}}{2}, \frac{-1 - \sqrt{5}}{2}$
29. $\frac{-1 + \sqrt{33}}{4}, \frac{-1 - \sqrt{33}}{4}$
30. $\frac{-3 + \sqrt{13}}{2}, \frac{-3 - \sqrt{13}}{2}$ 31. 2, -2
32. $-4, -5$ 33. 0, -2
34. $\frac{-1 + \sqrt{21}}{2}, \frac{-1 - \sqrt{21}}{2}$ 35. $-\frac{4}{3}, 2$
36. $\frac{-1 + \sqrt{31}}{3}, \frac{-1 - \sqrt{31}}{3}$
37. base = 4 cm, height = 6 cm
38. 20 ft by 8 ft 39. 1 second, 2 seconds

Practice B

1. $(x - 1)^2$ 2. $(x + 12)^2$ 3. $(x + 7)^2$
4. $\left(x + \frac{2}{3}\right)^2$ 5. $\left(x - \frac{1}{2}\right)^2$ 6. $(x - 6)^2$ 7. 9
8. 16 9. 121 10. $\frac{1}{16}$ 11. 81 12. 225
13. $\frac{81}{4}$ 14. $\frac{49}{4}$ 15. $\frac{1}{25}$ 16. 0.64 17. 144
18. $\frac{1}{36}$ 19. $-3, -4$ 20. 4, 1 21. 1, -15
22. $12 + 3\sqrt{15}, 12 - 3\sqrt{15}$
23. $4 + 2\sqrt{5}, 4 - 2\sqrt{5}$
24. $5 + 4\sqrt{3}, 5 - 4\sqrt{3}$
25. $\frac{5 + \sqrt{17}}{4}, \frac{5 - \sqrt{17}}{4}$
26. $\frac{-5 + 2\sqrt{15}}{5}, \frac{-5 - 2\sqrt{15}}{5}$
27. $\frac{7 + \sqrt{85}}{6}, \frac{7 - \sqrt{85}}{6}$ 28. $\frac{3}{2}, -\frac{1}{3}$
29. $\frac{1 + \sqrt{3}}{3}, \frac{1 - \sqrt{3}}{3}$
30. $\frac{1 + \sqrt{6}}{2}, \frac{1 - \sqrt{6}}{2}$ 31. $-2 + \sqrt{5}, -2 - \sqrt{5}$
32. 0, -4 33. 1, -1 34. $1 + \sqrt{2}, 1 - \sqrt{2}$
35. $\frac{-3 + \sqrt{17}}{2}, \frac{-3 - \sqrt{17}}{2}$
36. $\frac{-1 + \sqrt{29}}{2}, \frac{-1 - \sqrt{29}}{2}$
37. $\frac{-1 + \sqrt{5}}{2}, \frac{-1 - \sqrt{5}}{2}$ 38. $\frac{4}{3}, -\frac{5}{2}$
39. $\frac{-1 + \sqrt{85}}{14}, \frac{-1 - \sqrt{85}}{14}$ 40. 4.43 sec
41. length: 18 cm, width: 10 cm
42. 272 mi by 383 mi

Practice C

1. $(x - 3)^2$ 2. $(x + 13)^2$ 3. $(x - 9)^2$
4. $\left(x + \frac{3}{5}\right)^2$ 5. $\left(x + \frac{1}{2}\right)^2$ 6. $(7 - 2x)^2$ 7. 64
8. 81 9. $\frac{169}{4}$ 10. $\frac{9}{64}$ 11. $\frac{121}{4}$ 12. 400
13. $\frac{1}{81}$ 14. $\frac{225}{4}$ 15. $\frac{9}{100}$ 16. $-1, -5$
17. $-9, 1$ 18. 3 19. $-2 + \sqrt{11}, -2 - \sqrt{11}$
20. $-3 + \sqrt{10}, -3 - \sqrt{10}$
21. $3 + \sqrt{2}, 3 - \sqrt{2}$
22. $-4 + \sqrt{3}, -4 - \sqrt{3}$
23. $\frac{3 + \sqrt{5}}{2}, \frac{3 - \sqrt{5}}{2}$ 24. $\frac{1 + \sqrt{17}}{2}, \frac{1 - \sqrt{17}}{2}$

Lesson 12.4 *continued*

25. $\dfrac{3 + 2\sqrt{3}}{3}, \dfrac{3 - 2\sqrt{3}}{3}$

26. $\dfrac{2 + \sqrt{14}}{2}, \dfrac{2 - \sqrt{14}}{2}$

27. $\dfrac{-3 + \sqrt{37}}{4}, \dfrac{-3 - \sqrt{37}}{4}$

28. $\dfrac{4 + \sqrt{46}}{5}, \dfrac{4 - \sqrt{46}}{5}$

29. $\dfrac{3 + 2\sqrt{6}}{3}, \dfrac{3 - 2\sqrt{6}}{3}$

30. $\dfrac{-5 + \sqrt{33}}{4}, \dfrac{-5 - \sqrt{33}}{4}$ **31.** $4\sqrt{3}, -4\sqrt{3}$

32. $\dfrac{7 + 2\sqrt{7}}{7}, \dfrac{7 - 2\sqrt{7}}{7}$ **33.** $\dfrac{3}{4}, \dfrac{4}{3}$ **34.** $\dfrac{2}{3}, -4$

35. $3 + \sqrt{11}, 3 - \sqrt{11}$

36. $\dfrac{3 + \sqrt{15}}{3}, \dfrac{3 - \sqrt{15}}{3}$ **37.** $\dfrac{2}{3}, -1$

38. $\dfrac{5 + \sqrt{33}}{4}, \dfrac{5 - \sqrt{33}}{4}$ **39.** $0, \dfrac{12}{7}$

40. 12.5 sec **41.** 0.54 sec or 7.46 sec

42. 33 mi/h

Reteaching with Practice

1. $2, -\dfrac{1}{2}$ **2.** $-1, -\dfrac{1}{3}$ **3.** $-\dfrac{1}{2}, -\dfrac{3}{2}$

4–6. Sample answers are given. **4.** finding square roots because the equation is of the form $ax^2 + c = 0$ **5.** factoring because the quadratic can be easily factored **6.** using quadratic formula because the quadratic contains decimals.

Interdisciplinary Application

1. 0, 8 **2.**

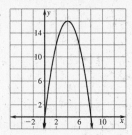

3. Answers vary. **4.** 2, 4 **5.** Answers vary.

Challenge: Skills and Applications

1. $4 \pm \sqrt{x + 16}$ **2.** $c < -16$

3, 5, 7. accept equivalent forms of solutions.

3. $\dfrac{5}{6} \pm \sqrt{c + \dfrac{25}{36}}$ **4.** $c < -\dfrac{25}{36}$

5. $\dfrac{3}{a} \pm \sqrt{\dfrac{4a + 9}{a^2}}$ for $a \neq 0 \left(\text{or } -\dfrac{2}{3} \text{ if } a = 0\right)$

6. $a < -\dfrac{9}{4}$

7. $\dfrac{b}{2} \pm \sqrt{-7 + \dfrac{b^2}{4}}$, or $\dfrac{b \pm \sqrt{b^2 - 28}}{2}$

8. $-2\sqrt{7} < b < 2\sqrt{7}$

9. $y = -16\left(x - \dfrac{1}{4}\right)^2 + 17$

10. $\dfrac{1}{4}$; it takes the pole vaulter $\dfrac{1}{4}$ second to reach the highest point in the jump.

11. 17; the pole vaulter reaches a height of 17 feet.

Lesson 12.5

Warm-up Exercises

1. 15 **2.** 41.2 **3.** 20 **4.** 13.6

Daily Homework Quiz

1. 64 **2.** $10 \pm \sqrt{102}$ **3.** $\dfrac{1 \pm \sqrt{5}}{4}$

4. about 6.8 cm and 8.8 cm

Lesson Opener

Allow 10 minutes.

1. 36; 64; 100; 100; they are equal.

2. 25; 144; 169; 169; they are equal.

3. 64; 225; 289; 289; they are equal.

4. 225; 400; 625; 625; they are equal.

5. 49; 576; 625; 625; they are equal.

6. 144; 1225; 1369; 1369; they are equal.

7. The sum of squares of the lengths of the two shorter sides equals the square of the length of the longest side.

Lesson 12.5 *continued*

Graphing Calculator Activity

1. yes **2.** no **3.** no **4.** no **5.** yes **6.** no

Practice A

1. $x, y; z$ **2.** $n, m; p$ **3.** $t, c; r$ **4.** $\sqrt{13}$

5. $\sqrt{61}$ **6.** $2\sqrt{34}$ **7.** $2\sqrt{13}$ **8.** $\sqrt{58}$

9. $6\sqrt{2}$ **10.** 15 **11.** 10 **12.** 13 **13.** 6, 8

14. 6, 6 **15.** 4, 8

16. No; $2^2 + 2^2 = 8$ and $4^2 = 16$

17. No; $6^2 + 9^2 = 117$ and $12^2 = 144$

18. No; $15^2 + 10^2 = 325$ and $20^2 = 400$

19. Yes; $10^2 + 24^2 = 676 = 26^2$

20. Yes; $5^2 + 5^2 = 50 = \left(5\sqrt{2}\right)^2$

21. Yes; $30^2 + 40^2 = 2500 = 50^2$

22. about 127.3 ft **23.** about 72.1 ft

Practice B

1. $\sqrt{73}$ **2.** $\sqrt{85}$ **3.** $\sqrt{89}$ **4.** $\sqrt{85}$

5. $4\sqrt{13}$ **6.** $5\sqrt{5}$ **7.** $\sqrt{290}$ **8.** 26

9. $3\sqrt{61}$ **10.** 9, 12 **11.** 9, 9

12. $\dfrac{7}{3}\sqrt{15}, \dfrac{14}{3}\sqrt{15}$

13. No; $4^2 + 4^2 = 32$ and $8^2 = 64$

14. Yes; $9^2 + 12^2 = 225 = 15^2$

15. No; $12^2 + 18^2 = 468$ and $24^2 = 576$

16. Yes; $0.3^2 + 0.4^2 = 0.25 = 0.5^2$

17. Yes; $8^2 + 8^2 = 128 = \left(8\sqrt{2}\right)^2$

18. Yes; $\left(\frac{3}{2}\right)^2 + \left(\frac{4}{2}\right)^2 = \frac{25}{4} = \left(\frac{5}{2}\right)^2$

19. hypothesis: a quadrilateral is a square; conclusion: it is a rectangle **20.** hypothesis: a triangle is equilateral; conclusion: all the side lengths are congruent **21.** hypothesis: today is February 29; conclusion: it is a leap year

22. about 40.5 feet **23.** about 8.9 feet

Practice C

1. $\sqrt{73}$ **2.** $\sqrt{106}$ **3.** $9\sqrt{2}$ **4.** $4\sqrt{13}$

5. $\sqrt{265}$ **6.** 25 **7.** 3 **8.** $\frac{5}{4}$ **9.** 1 **10.** 12, 9

11. 12, 12 **12.** 3, 6

13. No; $6^2 + 6^2 = 72$ and $10^2 = 100$

14. Yes; $12^2 + 16^2 = 400 = 20^2$

15. No; $15^2 + 18^2 = 549$ and $21^2 = 441$

16. No; $0.1^2 + 0.2^2 = 0.05$ and $0.3^2 = 0.09$

17. Yes; $12^2 + 12^2 = 288 = \left(12\sqrt{2}\right)^2$

18. Yes; $\left(\frac{6}{5}\right)^2 + \left(\frac{8}{5}\right)^2 = 4 = 2^2$

19. hypothesis: you visit the school's website; conclusion: you will see a picture of the math team **20.** hypothesis: all of the angles of a triangle measure less than 90°; conclusion: it is an acute triangle

21. hypothesis: the sides of a triangle measure 6 cm, 8 cm, and 12 cm; conclusion: the triangle is scalene

22. about 7.8 mi **23.** about 112.6 ft

24. about 116.6 ft

Reteaching with Practice

1. 15 **2.** $\sqrt{29}$ **3.** 36, 48

4. The lengths are sides of a right triangle.

5. The lengths are not sides of a right triangle.

6. The lengths are sides of a right triangle.

Real-Life Application

1. 90 feet **2.** about 64 feet **3.** about 36 feet

4. about 115 feet

Math and History

1. 13 feet and $10\frac{2}{3}$ inches **2.** about 447 meters

Challenge: Skills and Applications

1. $x = 3; AB = 25, BC = 15, AC = 20$

2. $x = 12; AB = 169, BC = 65, AC = 156$

3. about 17.0 in., about 16.97 in.

4. AE should equal BD, but BD is not quite 17 in. **5.** $d^2 = a^2 + b^2$ **6.** $e^2 = d^2 + c^2$

7. $e^2 = a^2 + b^2 + c^2$ **8.** 2 **9.** $\sqrt{11}$

10. $\sqrt{n+1}$

Quiz 2

1. 9 **2.** $1, -13$ **3.** 10.5 m **4.** no, the sum of the squares of two sides does not equal the square of the third side **5.** 20 m

Lesson 12.6

Warm-up Exercises

1. Celia's

Daily Homework Quiz

1. a. $b = \sqrt{481}$ **b.** $a = 3.35, b = 1.35$

2. No, $8^2 + 12^2 \neq 17^2$ **3.** hypothesis: a right triangle has sides that are 3 inches and 4 inches long; conclusion: the hypotenuse is 5 inches long

Lesson Opener

Allow 10 minutes.

1. Use the Pythagorean theorem.

2. 15 miles   **3.** Use the Pythagorean theorem.

4. 13 miles   **5.** Use the Pythagorean theorem.

6. 17 miles

Practice A

1–3. Sample estimates are given.   **1.** 8, 7.81

2. 6, 6.40   **3.** 7, 7.62

4. 4.24   **5.** 5   **6.** 3.61   **7.** 6.40   **8.** 10

9. 7.07   **10.** 13.45   **11.** 15   **12.** 4.24

13. yes   **14.** yes   **15.** no   **16.** no   **17.** (3, 3)

18. (3, 4)   **19.** (4, −2)   **20.** (−1, 2)

21. $\left(\frac{3}{2}, -\frac{7}{2}\right)$ **22.** $\left(\frac{3}{2}, \frac{3}{2}\right)$ **23.** $\left(-3, -\frac{3}{2}\right)$

24. $\left(0, -\frac{3}{2}\right)$ **25.** (1, 1)   **26.** about 24.33 km

27. about 39.6 km   **28.** about 44.72 km

Practice B

1. 5   **2.** 13   **3.** 2.24   **4.** 7.07   **5.** 10.05

6. 9.06   **7.** 12.53   **8.** 1.41   **9.** 3.61   **10.** 1.89

11. 3.05   **12.** 1.5   **13.** yes   **14.** no   **15.** yes

16. no   **17.** no   **18.** yes   **19.** (3, 7)

20. (5, −2)   **21.** (−2, −5)   **22.** (6, −4)

23. (−3, 6)   **24.** $\left(-1, \frac{5}{2}\right)$ **25.** $\left(-4, \frac{3}{2}\right)$

26. $\left(\frac{5}{2}, -1\right)$ **27.** $\left(-\frac{3}{2}, -\frac{1}{2}\right)$ **28.** $\left(\frac{5}{2}, \frac{3}{2}\right)$

29. $\left(-\frac{5}{2}, \frac{9}{2}\right)$ **30.** $\left(-\frac{3}{2}, -\frac{11}{2}\right)$ **31.** about 60.83 ft

32. about 288 mi   **33.** $725,000

34. $\left(\frac{11}{2}, \frac{3}{2}\right), \left(\frac{11}{2}, \frac{3}{2}\right)$; the midpoints of the diagonals are the same.

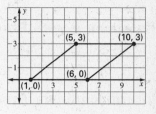

Practice C

1. 5.66   **2.** 4.12   **3.** 7.62   **4.** 6.71   **5.** 9.49

6. 11.18   **7.** 12.04   **8.** 5.66   **9.** 6.66

10. 6.93   **11.** 1.41   **12.** 1.41   **13.** yes   **14.** no

15. yes   **16.** yes   **17.** yes   **18.** no

19. (−1, 2)   **20.** $\left(\frac{7}{2}, -5\right)$ **21.** (5, −1)

22. (8, −4)   **23.** $\left(-3, -\frac{13}{2}\right)$ **24.** (0.7, 5)

25. (−1.9, 3)   **26.** (0, 1.9)   **27.** (−6, −3)

28. $\left(1\frac{3}{8}, 2\right)$ **29.** $\left(-\frac{3}{4}, 2\right)$ **30.** $\left(2\frac{2}{5}, -6\right)$

31. $AB = BC = CD = DA \approx 4.472$

32. midpoint of $\overline{AB} = (5, 2)$
midpoint of $\overline{BC} = (2, -1)$
midpoint of $\overline{CD} = (1, 0)$
midpoint of $\overline{DA} = (4, 3)$

33.

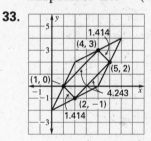

34. Perimeter of $\square ABCD \approx 17.888$
Perimeter of new quadrilateral ≈ 11.314

35. Midpoints are each (3, 1)

36.

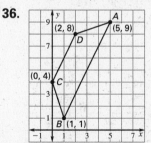

37. Slope $\overline{AB}$ = Slope $\overline{CD} = 2$

38. $BC = AD \approx 3.162$

Lesson 12.6 *continued*

Reteaching with Practice

1. 5 **2.** 2.83 **3.** 5.39 **4.** 17 mi **5.** $\left(\frac{5}{2}, 4\right)$

6. $(1, 0)$ **7.** $(3, 1)$ **8.** $(15, 25)$

Interdisciplinary Application

1. about 34 ft

2.

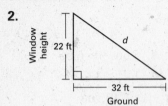

3. about 39 ft

4. about 72 ft **5.** The new wing design caused the plane to travel further.

Challenge: Skills and Applications

1. -3 **2.** $\frac{3}{5}$ **3.** $-5, 13$ **4.** $-2, 12$

5. $-9, 6$ **6.** $a = 3, b = -9$

7. $a = 4, b = -6$ **8.** $a = 4, b = 2$

9. $(2t, 0); (0, -8(t - 1))$

10. $2\sqrt{17t^2 - 32t + 16}$ **11.** about 8.9 mi

12. 1:30 P.M.

Lesson 12.7

Warm-up Exercises

1. b **2.** 33.73 **3.** 183.66

Daily Homework Quiz

1. 7.6 **2.** Yes,   **3.** $\left(-\frac{5}{2}, 5\right)$

4. 2.5 miles

Lesson Opener

Allow 10 minutes.

1–4.

	a	b	c	$\dfrac{a}{b}$	$\dfrac{a}{c}$	$\dfrac{b}{c}$
1.	34	59	68	0.576	0.5	0.868
2.	15	26	30	0.577	0.5	0.867
3.	22	38	44	0.579	0.5	0.864
4.	26	45	52	0.578	0.5	0.865

5. The angle measures are the same; the lengths of the sides and the orientations of the triangles are different.

6. The values in each column are equal or approximately equal.

Graphing Calculator Activity

1. $\frac{4}{5}$ **2.** $\frac{3}{5}$ **3.** $\frac{4}{3}$ **4.** $\cos 45°$ **5.** $\sin 55°$

6. $\cos 40°$ **7.** The tangent of complementary angles are reciprocals of one another.

Practice A

1. $\overline{PQ}$ **2.** $\overline{MP}$ **3.** $\overline{MP}$ **4.** $\overline{QP}$ **5.** $\overline{MQ}$

6. $\angle M, \angle Q$

7. $\sin A = \frac{4}{5}, \cos A = \frac{3}{5}, \tan A = \frac{4}{3};$
$\sin B = \frac{3}{5}, \cos B = \frac{4}{5}, \tan B = \frac{3}{4}$

8. $\sin A = \frac{12}{13}, \cos A = \frac{5}{13}, \tan A = \frac{12}{5};$
$\sin B = \frac{5}{13}, \cos B = \frac{12}{13}, \tan B = \frac{5}{12}$

9. $\sin A = \frac{21}{29}, \cos A = \frac{20}{29}, \tan A = \frac{21}{20};$
$\sin B = \frac{20}{29}, \cos B = \frac{21}{29}, \tan B = \frac{20}{21}$

10. $t \approx 16.96, r \approx 10.60$

11. $j \approx 10.28, l \approx 12.26$

12. $d \approx 12.74, e \approx 31.52$ **13.** about 13.77 ft

14. about 76.71 ft

Practice B

1. $\sin A = \frac{4}{5}, \cos A = \frac{3}{5}, \tan A = \frac{4}{3};$
$\sin B = \frac{3}{5}, \cos B = \frac{4}{5}, \tan B = \frac{3}{4}$

2. $\sin A = \frac{7}{25}, \cos A = \frac{24}{25}, \tan A = \frac{7}{24};$
$\sin B = \frac{24}{25}, \cos B = \frac{7}{25}, \tan B = \frac{24}{7}$

3. $\sin A = \frac{15}{17}, \cos A = \frac{8}{17}, \tan A = \frac{15}{8};$
$\sin B = \frac{8}{17}, \cos B = \frac{15}{17}, \tan B = \frac{8}{15}$

4. $d \approx 9.19, f \approx 7.71$ **5.** $x \approx 14.14, z \approx 14.14$

6. $q = 18, p \approx 15.59$ **7.** $k \approx 26.90, r \approx 19.99$

8. $d \approx 29.00, t \approx 13.52$

9. $h \approx 10.07, j \approx 15.66$ **10.** about 9.9 ft

11. about 50 ft

Practice C

1. $\sin A = \frac{9}{41}; \cos A = \frac{40}{41}, \tan A = \frac{9}{40};$
$\sin B = \frac{40}{41}, \cos B = \frac{9}{41}, \tan B = \frac{40}{9}$

Lesson 12.7 continued

2. $\sin A = \frac{15}{17}$, $\cos A = \frac{8}{17}$, $\tan A = \frac{15}{8}$;

$\sin B = \frac{8}{17}$, $\cos B = \frac{15}{17}$, $\tan B = \frac{8}{15}$

3. $\sin A = \frac{15}{17}$, $\cos A = \frac{8}{17}$, $\tan A = \frac{15}{8}$;

$\sin B = \frac{8}{17}$, $\cos B = \frac{15}{17}$, $\tan B = \frac{8}{15}$

4. $m \approx 4.64$, $e \approx 14.27$ **5.** $i \approx 7.62$, $h \approx 5.75$

6. $h \approx 14.34$, $j \approx 6.06$ **7.** $e \approx 33.36$, $f \approx 23.18$

8. $q \approx 28.30$, $n \approx 17.42$

9. $c \approx 16.43$, $n \approx 30.90$ **10.** about 741 mi; about 894 mi **11.** from Atlanta about 586 mi; from Memphis about 480 mi

Reteaching with Practice

1. $\sin A = \frac{4}{5}$, $\cos A = \frac{3}{5}$, $\tan A = \frac{4}{3}$, $\sin C = \frac{3}{5}$,

$\cos C = \frac{4}{5}$, $\tan C = \frac{3}{4}$ **2.** $\sin A = \frac{24}{25}$, $\cos A = \frac{7}{25}$,

$\tan A = \frac{24}{7}$, $\sin C = \frac{7}{25}$, $\cos C = \frac{24}{25}$, $\tan C = \frac{7}{24}$

3. $\sin A = \frac{40}{41}$, $\cos A = \frac{9}{41}$, $\tan A = \frac{40}{9}$,

$\sin C = \frac{9}{41}$, $\cos C = \frac{40}{41}$, $\tan C = \frac{9}{40}$

4. $a \approx 11.57$, $b \approx 13.79$

5. $b \approx 30.22$, $c \approx 32.16$

6. $a \approx 3.64$, $c \approx 10.64$

Cooperative Learning Activity

1. Answers will vary. **2.** *Sample answer:* This method of measurement might be useful in choosing a ladder of appropriate height. Foresters might also use this method to estimate the height of tree to be felled.

Real-Life Application

1. about 18 inches **2.** about 37 inches

3. $a^2 + b^2 = c^2$ **4.** about 21 inches

$9^2 + 36^2 = c^2$

$81 + 1296 = c^2$

$1377 = c^2$

$37.108 = c$

5. farther; about 6 inches

Challenge: Skills and Applications

1. $45°$ **2.** $a\sqrt{2}$ **3. a.** $\frac{\sqrt{2}}{2}$ **b.** $\frac{\sqrt{2}}{2}$ **c.** 1

---

4. a. $\frac{\sqrt{2}}{2} \approx 0.7071$ and $\sin 45° \approx 0.7071$, so checks **b.** $\frac{\sqrt{2}}{2} \approx 0.7071$ and $\cos 45° \approx 0.7071$, so checks **c.** $\tan 45° = 1$, so checks

5. a. 0.5 **b.** 0.8660 **c.** 0.5774 **d.** 0.5774

e. $\frac{\sin 30°}{\cos 30°} = \tan 30°$

f. $\dfrac{\sin A}{\cos A} = \dfrac{\dfrac{\text{side opposite } A}{\text{hypotenuse}}}{\dfrac{\text{side adjacent to } A}{\text{hypotenuse}}} =$

$\dfrac{\text{side opposite } A}{\text{side adjacent to } A} = \tan A$

6. 8.00 to nearest hundredth, so about 8 ft

7. about 3.38 ft **8.** 21.70 to nearest hundredth, so about 21.7 ft

Lesson 12.8

Warm-up Exercises

1. distributive property **2.** addition property of equality **3.** commutative property of addition

4. identity property of addition

5. inverse property of multiplication

Daily Homework Quiz

1. $\sin A = \frac{7}{25}$; $\sin B = \frac{24}{25}$;

$\cos A = \frac{24}{25}$; $\cos B = \frac{7}{25}$;

$\tan A = \frac{7}{24}$; $\tan B = \frac{24}{7}$

2. $y = 9.83$, $x = 8.98$ **3.** 27.2 ft

Lesson Opener

Allow 10 minutes.

1. E **2.** A **3.** D **4.** C **5.** F **6.** B

Practice A

1. Identity prop. of add. **2.** Dist. prop.

3. Commut. prop. of mult.

4. Identity prop. of mult.

5. Commut. prop. of add.

6. Inverse prop. of add.

Lesson 12.8 *continued*

7. Assoc. prop. of mult.

8. Inverse prop. of mult.

9. Assoc. prop. of add.

10. **a.** Add. axiom of equality

    b. Given

    c. Add. axiom of equality

11. **a.** Identity prop. of mult.

    b. Inverse prop. of mult.

12. *Sample answer:* $|3 + (-3)| \neq |3| + |-3|$

13. *Sample answer:* $\sqrt{9 + 16} \neq \sqrt{9} + \sqrt{16}$

14. $AC = \sqrt{(x - 0)^2 + (y - 0)^2} = \sqrt{x^2 + y^2}$
    and
    $BD = \sqrt{(0 - x)^2 + (y - 0)^2} = \sqrt{x^2 + y^2}$,
    so $AC = BD$.

Practice B

1. Commut. prop. of add.

2. Inverse prop. of add.   **3.** Dist. prop.

4. Assoc. prop. of mult.

5. Commut. prop. of mult.

6. Identity prop. of add.

7. Identity prop. of mult.

8. Inverse prop. of mult.

9. Assoc. prop. of add.

10. **a.** Add. axiom of equality   **b.** Given

    c. Add. axiom of equality   **d.** Def. of subtr.

11. $p - q = p + (-q)$ Def. of subtr.

    $p - q = -q + p$ Commut. prop. of add.

12. $(a + b) + c = a + (b + c)$
          Assoc. prop. of add.

    $a + (b + c) = a + (c + b)$
          Commut. prop. of add.

    $(a + b) + c = a + (c + b)$
          Subst. prop. of equality

    $a + (c + b) = (a + c) + b$
          Assoc. prop. of add.

    $(a + b) + c = (a + c) + b$
          Subst. prop. of equality

13. $(m + n) - n = (m + n) + (-n)$
          Definition of subtraction

    $(m + n) - n = m + [n + (-n)]$
          Assoc. prop. of add.

$(m + n) - n = m + 0$ Inverse prop. of add.

$(m + n) - n = m$ Identity prop. of add.

14. *Sample answer:* $|-3 - 3| \neq |-3| - |3|$

15. *Sample answer:* $\sqrt{25 - 9} \neq \sqrt{25} - \sqrt{9}$

16. $AC = \sqrt{(x - 0)^2 + (y - 0)^2} = \sqrt{x^2 + y^2}$
    and $BD = \sqrt{(0 - x)^2 + (y - 0)^2} = \sqrt{x^2 + y^2}$ so
    $AC = BD$.

17. midpoint of $\overline{AC} = \left(\dfrac{x + 0}{2}, \dfrac{y + 0}{2} \right) = \left(\dfrac{x}{2}, \dfrac{y}{2} \right)$

    and midpoint of $\overline{BD} = \left(\dfrac{0 + x}{2}, \dfrac{y + 0}{2} \right) = \left(\dfrac{x}{2}, \dfrac{y}{2} \right)$,

    so the midpoints are the same. Thus, the diagonals of any rectangle bisect each other.

Practice C

1. Commut. prop. of add.

2. Inverse prop. of add.   **3.** Dist. prop.

4. Assoc. prop. of mult.

5. Commut. prop. of mult.

6. Identity prop. of add.

7. Identity prop. of mult.

8. Inverse prop. of mult.

9. Assoc. prop. of add.

10. **a.** Given

    b. Add. axiom of equality

    c. Given

    d. Add. axiom of equality

    e. Substitution prop. of equality

    f. Def. of subtr.

11. $a - b = a + (-b)$ Def. of subtr.

    $a - b = -b + a$ Commut. prop. of add.

12. $a + c = b + c$ Given

    $(a + c) + (-c) = (b + c) + (-c)$
          Add. axiom of equality

    $a + [c + (-c)] = b + [c + (-c)]$
          Assoc. prop. of add.

    $a + 0 = b + 0$ Inverse prop. of add.

    $a = b$ Identity prop. of add.

13. $(b + a) - b = (a + b) - b$
          Commut. prop. of add.

    $(b + a) - b = (a + b) + (-b)$ Def. of subtr.

    $(b + a) - b = a + [b + (-b)]$
          Assoc. prop. of add.

Lesson 12.8 *continued*

$(b + a) - b = a + 0$ Inverse prop. of add.

$(b + a) - b = a$ Identity prop. of add.

14. *Sample Answer:* $-3 < 2$ but $(-3)(-2) \not< 2(-2)$

15. *Sample Answer:* $-3 < 2$ but $(-3)^2 \not< 2^2$

16. $AC = \sqrt{[(x - b) - 0]^2 + (c - 0)^2}$

$= \sqrt{(x - b)^2 + c^2}$

$= \sqrt{x^2 - 2bx + b^2 + c^2}$

$BD = \sqrt{(b - x)^2 + (c - 0)^2}$

$= \sqrt{b^2 - 2bx + x^2 + c^2}$, so the lengths of the diagonals are equal.

Reteaching with Practice

1.

$ac = bc, c \neq 0$	Given
$ac\left(\dfrac{1}{c}\right) = bc\left(\dfrac{1}{c}\right)$	Multiplication axiom of equality
$a\left(c \cdot \dfrac{1}{c}\right) = b\left(c \cdot \dfrac{1}{c}\right)$	Associative axiom of multiplication
$a(1) = b(1)$	Inverse axiom of mult.
$a = b$	Identity axiom of mult.

2.

$a - c = b - c$	Given
$a + (-c) = b + (-c)$	Def. of Subtraction
$a + (-c) + c = b + (-c) + c$	Addition axiom of equality
$a + c + (-c) = b + c + (-c)$	Comm. axiom of add.
$a + [c + (-c)] = b + [c + (-c)]$	Associative axiom of addition
$a + 0 = b + 0$	Inverse axiom of add.
$a = b$	Identity axiom of add.

3. Answers vary. For example, let $a = 3$ and $b = 4$. $\sqrt{a^2 + b^2} = \sqrt{9 + 16} = 5$ but $a + b = 7$. $\sqrt{a^2 + b^2} \neq a + b$

4. Answers vary. For example, let $a = 3$ and $b = 4$. $a - b = 3 - 4 = -1$ but $b - a = 1$. $a - b \neq b - a$

5. Answers vary. For example, let $a = 2$ and $b = 4$. $a \div b = 2 \div 4 = 0.5$ but $b \div a = 4 \div 2 = 2$. $a \div b \neq b \div a$

6. If $a + b > b + c$, then $a > c$. Assume that $a > c$ is not true. Then $a \leq c$.

$a \leq c$	Assume the opposite of $a > c$ is true.
$a + b \leq c + b$	Adding the same number to each side produces an equivalent inequality.

This contradicts the given statement that $a + b > b + c$. Therefore, it is impossible that $a \leq c$. You conclude that $a > c$ is true.

Interdisciplinary Application

1. Eratosthenes uses this theorem to show that the angle formed by the tower and the sun's rays is equal to the angle formed by the well and the tower at the center of the Earth.

2. This Theorem is necessary because he could not physically measure both of these angles. He could only measure one. Using the Theorem, he could prove the angle he could not measure. (This is the key to his success.)

Challenge: Skills and Applications

1–3, 4c. Check students' sequences of steps. Sample proofs are given.

1. $23 + 45$

$= (20 + 3) + (40 + 5)$ write 23 and 45 as sums

$= 20 + [3 + (40 + 5)]$ assoc. prop. of add.

$= 20 + [(3 + 40) + 5]$ assoc. prop. of add.

$= 20 + [(40 + 3) + 5]$ commut. prop. of add.

$= [20 + (40 + 3)] + 5$ assoc. prop. of add.

$= [(20 + 40) + 3] + 5$ assoc. prop. of add.

$= (20 + 40) + (3 + 5)$ assoc. prop. of add.

2. $(10a + b) + (10c + d)$

$= 10a + [b + (10c + d)]$ assoc. prop. of add.

$= 10a + [(b + 10c) + d]$ assoc. prop. of add.

$= 10a + [(10c + b) + d]$ commut. prop. of add.

$= [10a + (10c + b)] + d$ assoc. prop. of add.

$= [(10a + 10c) + b] + d$ assoc. prop. of add.

$= (10a + 10c) + (b + d)$ assoc. prop. of add.

3. $(10a + b) - (10c + d)$

$= (10a + b) + [-(10c + d)]$ def. of subtr.

$= (10a + b) + [-1(10c + d)]$ prop. of opp.

$= (10a + b) + [(-1)(10c) + (-1)d]$ dist. prop.

$= (10a + b) + [(-10c) + (-d)]$ prop. of opp.

$= 10a + \{b + [(-10c) + (-d)]\}$ assoc. prop.

$= 10a + \{[b + (-10c)] + (-d)\}$ assoc. prop.

$= 10a + [(-10c + b) + (-d)]$ commut. prop.

$= [10a + (-10c + b)] + (-d)$ assoc. prop.

$= \{[10a + (-10c)] + b\} + (-d)$ assoc. prop.

$= [10a + (-10c)] + [b + (-d)]$ assoc. prop.

$= (10a - 10c) + (b - d)$ def. of subtr.

4. a. 1035 **b.** $(2 \cdot 10 + 3)(4 \cdot 10 + 5)$;
1035; yes

c. $(10a + b)(10c + d)$

$= (10a + b)(10c) + (10a + b)d$ dist. prop.

$= (10a)(10c) + b(10c) + (10a)d + bd$ dist. prop.

$= (10a \cdot 10)c + (b \cdot 10)c + 10(ad) + bd$ assoc. prop. of mult.

$= [10 \cdot (10a)]c + (10 \cdot b)c + 10(ad) + bd$ commut. prop. of mult.

$= [(10 \cdot 10)a]c + 10(bc) + 10(ad) + bd$ assoc. prop. of mult.

$= 100(ac) + 10(bc) + 10(ad) + bd$ assoc. prop. of mult. and simplify

$= 100ac + 10(bc + ad) + bd$ dist. prop.

5. *Sample answer:* Guess 32. If this is too high, then the number is between 1 and 31, so guess 16. Otherwise, the number is between 33 and 64, so guess $32 + 16 = 48$. Keep guessing the middle number in the appropriate range or, if the number of values in the range is even, guess the lower of the two middle numbers. (For example, there are 32 numbers between 33 and 64 and the middle two numbers are 48 and 49, so you guess 48.)

6. 6 **7.** 7, n **8.** On the first guess, you reduce the possible number of correct integers to $2^{n-1} - 1$ or to 2^{n-1}, on the second guess to $2^{n-2} - 1$ or to 2^{n-2}, and so on. After $n - 1$ guesses, either the possible number of correct integers is $2^{n-(n-1)} - 1 = 2 - 1 = 1$ or it is $2^{n-(n-1)} = 2$. If it is 1, then you are done in $n - 1$ guesses. If it is 2, then you need one more guess. So, the correct integer must be guessed on or before the nth guess.

Review and Assessment

Review Games and Activities

1. $2\sqrt{15} - 8\sqrt{6}$ **2.** $15 - 2\sqrt{14}$

3. $9\sqrt{2}$ **4.** $x = 22$ **5.** $x = 23$ **6.** $c = \sqrt{13}$

7. $b = 4\sqrt{6}$ **8.** $\sqrt{274}$ **9.** $3\sqrt{5}$

KNOTTY PINES

Test A

1. domain: all nonnegative numbers; range: all nonnegative numbers

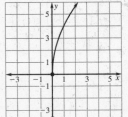

2. domain: all nonnegative numbers; range: all numbers greater than or equal to 2

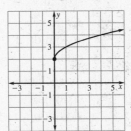

3. $7\sqrt{5}$ **4.** $4\sqrt{3}$ **5.** 10 **6.** $\dfrac{3\sqrt{2}}{2}$

7. $6\sqrt{10} - 5$ **8.** $\dfrac{2\sqrt{5}}{5}$ **9.** $x = 4$

10. $x = 12$ **11.** $a = 36$ **12.** $a = 27$

13. $x = -3$ or $x = 1$ **14.** $x = -4 \pm \sqrt{30}$

15. $c = 5$ **16.** $c = 15$ **17.** 5 **18.** 10

Review and Assessment *continued*

19. Yes   **20.** No

21. approximately 12.17 miles   **22.** (4, 3)

23. (3, 2)   **24.** \$610,000   **25.** $\sin R = \frac{4}{5}$;

$\cos R = \frac{3}{5}$; $\tan R = \frac{4}{3}$; $\sin S = \frac{3}{5}$; $\cos S = \frac{4}{5}$;

$\tan S = \frac{3}{4}$ **26.** $\sin R = \frac{4}{5}$; $\cos R = \frac{3}{5}$; $\tan R = \frac{4}{3}$;

$\sin S = \frac{3}{5}$; $\cos S = \frac{4}{5}$; $\tan S = \frac{3}{4}$ **27.** $x = 8$

28. $x \approx 11.31$

Test B

1. domain: all nonnegative numbers; range: all numbers greater than or equal to -2

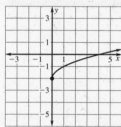

2. domain: all numbers greater than or equal to -1; range: all nonnegative numbers

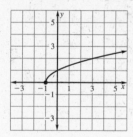

3. $4\sqrt{3}$ **4.** $2\sqrt{5}$ **5.** $4\sqrt{6} + 3$ **6.** $\frac{7\sqrt{5}}{5}$

7. 1   **8.** $\frac{48 + 6\sqrt{3}}{61}$ **9.** $x = 9$ **10.** $x = 36$

11. $a = 32$ **12.** $a = 32$ **13.** $x = -3 \pm \sqrt{13}$

14. $x = -2$ and $x = 1$ **15.** $b = 15$

16. $x = 4; 6, 8$ **17.** 5   **18.** $\sqrt{41}$ **19.** Yes

20. No   **21.** approximately 14 miles   **22.** $\left(\frac{3}{2}, \frac{5}{2}\right)$

23. $\left(-\frac{1}{2}, 2\right)$ **24.** \$637,500

25. $\sin R = \frac{28}{53}$; $\cos R = \frac{45}{53}$; $\tan R = \frac{28}{45}$;

$\sin S = \frac{45}{53}$; $\cos S = \frac{28}{53}$; $\tan S = \frac{45}{28}$

26. $\sin R = \frac{33}{65}$; $\cos R = \frac{56}{65}$; $\tan R = \frac{33}{56}$;

$\sin S = \frac{56}{65}$; $\cos S = \frac{33}{65}$; $\tan S = \frac{56}{33}$

27. $x \approx 14.43$ **28.** $x \approx 9.24$

Test C

1. domain: all numbers greater than or equal to -5; range: all nonnegative numbers

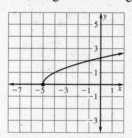

2. domain: all numbers greater than or equal to $-\frac{3}{2}$; range: all nonnegative numbers

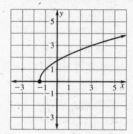

3. $5\sqrt{6}$ **4.** $-3\sqrt{7}$ **5.** $91 + 40\sqrt{3}$

6. $\frac{8(4 - \sqrt{5})}{11}$ **7.** $-29\sqrt{5} - 81$

8. $-\frac{11 + 4\sqrt{7}}{3}$ **9.** $x = 3$ **10.** $x = 12$

11. $a = 1536$ **12.** $a = 2916$

13. $x = \frac{1 \pm \sqrt{21}}{2}$ **14.** $x = \frac{-3 \pm \sqrt{41}}{4}$

15. $x = 2; 6, 8$ **16.** $x = 5; 9, 12$ **17.** 3

18. 10   **19.** Yes   **20.** No

21. approximately 12 miles   **22.** $\left(\frac{1}{2}, -\frac{1}{2}\right)$

23. $(-1, -3)$ **24.** \$3,412,500

25. $\sin R = \frac{88}{137}$; $\cos R = \frac{105}{137}$; $\tan R = \frac{88}{105}$;

$\sin S = \frac{105}{137}$; $\cos S = \frac{88}{137}$; $\tan S = \frac{105}{88}$

26. $\sin R = \frac{60}{109}$; $\cos R = \frac{91}{109}$; $\tan R = \frac{60}{91}$;

$\sin S = \frac{91}{109}$; $\cos S = \frac{60}{109}$; $\tan S = \frac{91}{60}$

Answers

27. $x \approx 13.86$ **28.** $x \approx 13.86$

SAT/ACT Chapter Test

1. B   **2.** D   **3.** C   **4.** B   **5.** D   **6.** A   **7.** A
8. A

Alternative Assessment

1. a–c. Complete answers should address these points.   **a.** • Explain that the graphs all shift up 2 units when 2 is added to the function, and explain how they all shift down when 1 is subtracted from the function.   • Explain that when 3 is added to the x-value in the function, the graphs are shifted 3 units to the left from the parent function.   **b.** • Explain that if h is positive, the graph will shift to the right h units, and if h is negative, the graph will shift to the left $|h|$ units. If k is positive, the graph will shift up k units, and if k is negative, the graph will shift down $|k|$ units from the parent graph. If either h or k is zero, the graph does not shift for that portion.   **c.** • Explain that a negative sign in front of the function will cause a reflection of the graph in the x-axis.

2. a. $(5, 0), (8, -3), (5, -6)$

b. $3\sqrt{2}, 3\sqrt{2}$, and 6

c. Yes, it is a right triangle because the sides fit the conditions of the Pythagorean theorem.

d. $6 + 6\sqrt{2}$ units; 9 square units

e. $\sin B = \cos B = \dfrac{\sqrt{2}}{2}$; $\tan B = 1$ **f.** 5, -11

3. $6\sqrt{2}, 6\sqrt{2}, 12$; The original triangle is a right triangle. *Sample answer:* Since the original triangle is a right triangle, the new triangle formed by connecting its midpoints must also be a right triangle because the triangles are similar triangles.

Project: Puzzling Distances

1. $(4, 3)$ **2.** connect $(4, 3)$ to $(2, 3)$

3. $(0, 5)$, connect $(2, 3)$ to $(0, 5)$

4. connect $(4, 3)$ to $(4, 1)$ to $(5, 0)$

Cumulative Review

1. $\frac{25}{36}$ **2.** 2   **3.** $\frac{31}{2}$ **4.** $\frac{86}{27}$ **5.** 1.16 cm

6. 3.27 m   **7.** no   **8.** no   **9.** yes

10. no   **11.** $y = \frac{1}{5}x + \frac{2}{5}$

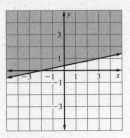

12. $y = -\frac{7}{9}x + \frac{16}{9}$

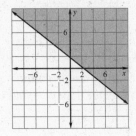

13. $y = \frac{7}{10}x + 7$

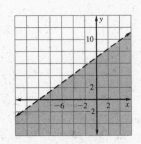

14. $10{,}837.50   **15.** $3{,}474.25   **16.** $1310.31

17. $\dfrac{\sqrt{15}}{6}$ **18.** $2\sqrt{2}$ **19.** $\dfrac{4\sqrt{10}}{7}$

20. $-12t(8t - 1)(t + 1)$ **21.** $3t^4(t - 8)(t + 1)$

22. x^2 **23.** $\dfrac{x^5}{x + 2}$ **24.** $x \geq 4$ **25.** $x \geq 0$

26. $x \geq -8$ **27.** $x \geq \frac{1}{9}$ **28.** $x \geq -\frac{1}{2}$

29. $x \neq 0, x \leq 1$ **30.** $23\sqrt{2}$ **31.** $9\sqrt{5}$

32. -3 **33.** $3\sqrt{6} + 10$ **34.** 1156   **35.** 12

36. 3   **37.** 2   **38.** $2 \pm \sqrt{2}$ **39.** $-\dfrac{5}{2} \pm \dfrac{\sqrt{57}}{2}$

40. $\dfrac{1}{3} \pm \dfrac{2\sqrt{7}}{3}$ **41.** $-\dfrac{2}{5} \pm \dfrac{\sqrt{29}}{5}$ **42.** 26 in.

43. 4 cm   **44.** $7\sqrt{3}$ m, 14 m   **45.** 11.18

46. 6.08   **47.** 18.44   **48.** 1.67   **49.** 6.93   **50.** 17.10

51. $\sin R = \frac{45}{53}$, $\cos R = \frac{28}{53}$, $\tan R = \frac{45}{28}$

52. $\sin R = \frac{25.9}{70.9}$, $\cos R = \frac{66}{70.9}$, $\tan R = \frac{25.9}{66}$

53. $\sin R = \frac{5.1}{14.9}$, $\cos R = \frac{14}{14.9}$, $\tan R = \frac{5.1}{14}$